Sun Protection

A risk management approach

Sun Protection

A risk management approach

Brian Diffey

Dermatological Sciences, Institute of Cellular Medicine, Newcastle University

IOP Publishing, Bristol, UK

ISBN   978-0-7503-1377-3 (ebook)
ISBN   978-0-7503-1378-0 (print)
ISBN   978-0-7503-1379-7 (mobi)

DOI   10.1088/978-0-7503-1377-3

Version: 20171001

IOP Expanding Physics
ISSN 2053-2563 (online)
ISSN 2054-7315 (print)

British Library Cataloguing-in-Publication Data: A catalogue record for this book is available from the British Library.

Published by IOP Publishing, wholly owned by The Institute of Physics, London

IOP Publishing, Temple Circus, Temple Way, Bristol, BS1 6HG, UK

US Office: IOP Publishing, Inc., 190 North Independence Mall West, Suite 601, Philadelphia, PA 19106, USA

Contents

Preface

On the journey from solar radiation entering the skin to eventually causing clinical effects, a number of pathways are involved that encompass a variety of seemingly unconnected areas of knowledge ranging from climatology, optical physics, photochemistry, cellular and molecular biology through to epidemiology, dermatology and pathology. Each of these subjects could be the topic of a book in its own right and there exist many excellent texts exploring one aspect of this pathway in detail.

This book deliberately does not compete with these texts but rather dips into the various stages along the journey and attempts to draw them together so that the expert is likely to learn little from the chapter dealing with his or her specialism but may find the complementary topics in other chapters of some interest. The book is written at a level that the generalist should be able to follow and so gain an overall picture of how we can apply the principles of risk management to enjoying and benefitting from sun exposure, whilst at the same time limiting our risk to acceptable levels.

Foreword

Over the last few days, since Professor Diffey kindly asked me to write this Foreword, I have been thinking about the significance of the timing of this invitation. Firstly, on a personal level, this month marks the start of my own thirtieth year within the sun protection industry. I will readily admit that I have lost not an ounce of excitement for the remarkably complete, elegant science that characterises this field, uniting the Physics of radiative transfer, the Biology of tissue response, and the Chemistry of sunscreens.

Secondly, on a global level, this past year has seen further recognition that skin cancer, linked inseparably to exposure to solar ultraviolet (UV) radiation, now constitutes the most common form of malignancy in humans. Furthermore, while thankfully only a minority will develop cancer, every human will nevertheless accumulate distinct deleterious changes in UV-exposed skin, serving as powerful perceptual cues for a loss of youth and health, with concomitant profound effects on self-worth and confidence.

The depth, breadth and longevity of Professor Diffey's contributions to this critical field of science are indisputable and unparalleled. Of equal significance, however, is the consistent manner in which the output of his research has been communicated—in a wonderfully clear, passionate, no-nonsense style which embraces the scientist and lay-person alike, whether in print, on platform or in conversation. Indeed, this present volume is no exception. Without a doubt, it represents the most comprehensive, authoritative and accessible text to yet cover this field and I recommend it to you warmly and without hesitation.

Professor Paul Matts
Research Fellow in Skin Care, Procter & Gamble
October 2017

Acknowledgments

I am indebted to the many individuals who kindly provided images and/or data for inclusion in this book, and to those colleagues who gave critical feedback on sections of the text specific to their respective expertise.

I hesitate to thank individuals by name for fear of omitting someone but will make one exception. My 40 year involvement with sun exposure and its effects in human skin would probably not have happened without the encouragement of the late Professor Ian Magnus. His wisdom and guidance were critical in my early development in the subject and for this I am forever grateful.

Author biography

Brian Diffey

Brian Diffey is Emeritus Professor in Dermatological Sciences at Newcastle University, UK. His career was spent in the NHS, where he was Professor of Medical Physics and Clinical Director in the Newcastle Hospitals. His involvement with sun protection has spanned more than 40 years and his interests include the measurement of personal sun exposure, its effects in normal and diseased skin, and ways to minimise excessive exposure, especially through the use of topical sunscreens.

He has advised a number of bodies on sun exposure and skin health including the World Health Organization, Department of Health, and the Cancer Research UK SunSmart programme, as well as patient support groups concerned with sun-related diseases such as vitiligo and xeroderma pigmentosum.

He invented both the UVA Star Rating for sunscreens in conjunction with Boots in the UK, and the Critical Wavelength adopted by the Food & Drug Administration in the USA as the sole measure of broad spectrum protection.

In 1999 he was awarded the Medal of the Society of Cosmetic Scientists for his contributions to suncare, and in 2011 was honoured at the International Sun Protection Conference for significant innovation in the field of photoprotection.

He is an honorary member of the British Association of Dermatologists, the Swedish Society for Dermatology and Venereology, and the European Society for Photodermatology.

IOP Publishing

Sun Protection
A risk management approach
Brian Diffey

Chapter 1

Introduction to risk management

1.1 Risk, hazard and exposure

There is adequate evidence that exposure to solar ultraviolet (UV) radiation is a major aetiological factor in human skin cancer but managing the risk of skin cancer does not necessarily mean avoiding exposure to the Sun's UV rays. For example, a young man who walks in the hills on a summer's day is choosing not to minimise his risk of skin cancer, but rather to face it and embrace it as part of an attempt to maximise his enjoyment and quality of life. A pragmatic approach, therefore, is to adopt strategies that control the *hazard*—that is, solar UV radiation—commensurate with the need or desire to be outdoors.

The terms *hazard, exposure* and *risk* have all been used in the opening paragraph so before proceeding it is probably worthwhile distinguishing between them.

A *hazard* exists where an object, substance or situation has an inherent ability to cause an adverse effect. Examples of hazards include uneven pavements, unguarded machinery, a fire, and, of course, UV radiation.

Exposure is the extent to which the likely recipient of the harm is exposed to, or can be influenced by, the hazard.

Risk, on the other hand, is the chance that adverse effects will occur; the risk can be high or negligible. For harm to occur—in other words, for there to be a risk—there must be BOTH the hazard AND the exposure to that hazard; without both these at the same time, there is no risk.

In terms of sun exposure there is an additional concept to consider—that of *vulnerability*, since two people experiencing the same UV exposure will not necessarily suffer the same health consequences. This is encapsulated by the *sun-reactive skin type*, which will be explained in chapter 4.

The control of our exposure to the Sun is termed *Sun Protection* or *Photoprotection* and this will be the theme of this book. A risk management approach, inspired by a proposal suggested by Haddon [1], will be adopted (section 1.4) to determine whether, how and in what circumstances, harm might be caused—the

doi:10.1088/978-0-7503-1377-3ch1

risk—and to explore the feasibility of various strategies in controlling exposure to solar ultraviolet radiation.

1.2 Risk management

Risk Management is the process leading to a choice among alternative courses of action and establishing the priorities and strategies for implementation. The process of risk management can be broadly split into *determination* and *evaluation*.

Determination involves identification of the hazard and estimation of the magnitude of the risk; this process is principally a *scientific* activity. Evaluation, on the other hand, looks at how different individuals accept the risk and what steps they might take to avert the risk; this processes is principally a *social* activity. A schematic diagram that identifies the various contributors to the risk management process, with particular reference to solar UV exposure, is shown in figure 1.1.

We believe solar UV radiation is detrimental to health from both observation and experimental studies. For example, skin cancers occur mainly on sites such as the hands and face that are chronically exposed to sunlight, and their incidence is greater in people living in sunny countries compared with people of similar susceptibility who live in less sunny locations. In addition, the results from experimental studies in model systems, such as the hairless mouse, show unequivocally that UV radiation induces skin cancers.

In estimating the magnitude of the risk we normally combine incidence data of a specific type of skin cancer from populations in different countries with data on ambient solar UV radiation at these geographical locations. We also observe how incidence changes with age and skin colour. We can then incorporate these epidemiological data into mathematical models of varying complexity that allow us to estimate the magnitude of risk for a defined set of inputs.

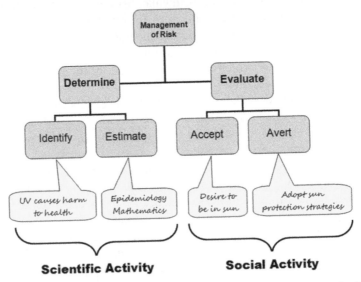

Figure 1.1. Contributors to the risk management process with particular reference to solar UV exposure.

People perceive risk differently depending on their awareness and understanding of the risk in question, and also on their perception of how they might suffer personally as a consequence of the risk. For example, those individuals who regard a tanned skin as an attractive, social attribute might well engage in sunbathing even though they are aware that sun exposure is a major risk factor for skin cancer. On the other hand, people with fair skin and a family history of skin cancer are more likely than others to adopt avoidance behaviour in strong sunlight.

1.3 Sun exposure and risk

Our relationship with the Sun is illustrated in figure 1.2.

Too little sun exposure leads to reduced well-being, notably low levels of vitamin D that are detrimental to our bone health. Furthermore, reduced plasma vitamin D levels have been reported to be associated with an increased risk of some cancers including colon and breast, as well as autoimmune diseases such as multiple sclerosis.

Many people believe that sunlight is good for their well-being, and there are a number of reports suggesting that sun exposure has beneficial systemic effects that include reduced blood pressure and a reduction in serum cholesterol. However, these and some other claims, remain controversial. There is a well-recognised clinical disorder called seasonal affective disorder (SAD), in which sufferers experience mood changes in the dark winter months. The length and severity of depression are generally influenced by latitude (milder nearer the equator) and weather (prolonged and worsened by cloudy days). The depressive episodes generally end in the spring.

The belief that sunlight is beneficial to health has existed for more than two millennia and persists to this day. Sunbathing was prescribed by many early physicians for the treatment of countless diseases and the practice continued into the 20th century (see section 2.3.2). Such was the enthusiasm for the healing powers of the Sun that in May 1924 the Sunlight League was formed in London and stated as one of its aims '...*the education of the public to the appreciation of sunlight as a*

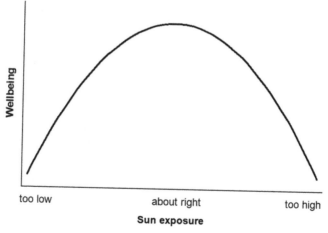

Figure 1.2. A sketch of the relationship between well-being and sun exposure.

means of health; teaching the nation that sunlight is Nature's universal disinfectant, as well as a stimulant and tonic'.

An enthusiastic protagonist of the role of sunlight in the proper care of the sick was Florence Nightingale. In her book *Notes on Nursing*, published in December 1859, she spoke of the 'acute suffering' caused to a sick person by being so placed that it is impossible to see out of the window.

A secondary benefit of sun exposure is that it encourages outdoor physical activity and this, of course, is beneficial to our health in terms of obesity, diabetes and coronary disease.

But we should not forget that too much sunlight on a summer day can lead that evening to skin that is red, painful and possibly blistering. And sun exposure over many years leads to skin cancer, the most common human cancer, with its associated morbidity and mortality.

So we need to strike a balance and aim for the peak of the well-being—sun exposure curve in figure 1.2. The purpose of this book is to show that with appropriate protection measures we can benefit from and enjoy the Sun, whilst at the same time reduce the adverse risks of sun exposure to what many people find acceptable.

1.4 A risk management approach to sun protection

In his seminal paper, Haddon [1] formulated a number of strategies to counter the damage resulting from exposure to a hazard and these strategies can be summarised below, with the chapters dealing with the specific strategy given in parentheses:

- Prevent the existence of the hazard or, if this is not possible, reduce the amount of hazard (chapters 2 and 3).
- Define the human impact of the hazard (chapter 4).
- Separate in space and time the hazard from individuals (chapter 5).
- Separate the hazard by the use of physical and chemical barriers (chapters 6 and 7).
- Make people more resistant to the hazard and begin to counter the damage already done (chapter 8).
- Stabilise, repair and rehabilitate the damage caused by the hazard (chapter 9).

Another way to summarise the contents of this book is to consider the so-called Haddon matrix [1], which is the most commonly used paradigm in addressing injury prevention. In its simplest form, the matrix has two dimensions.

The first dimension is based on the fact that the adverse clinical and societal end results of damaging interactions with solar UV radiation are preceded by processes that naturally divide into three stages. These three stages are the *pre-event*, *event*, and *post-event* phases, where *event* refers to the observable harm caused by the exposure.

The second dimension of the matrix is divided into the three factors—*human*, *intervention*, and *environment*—that influence the likelihood and magnitude of harm. Here, the term *environment* relates to both the physical and social environments (table 1.1).

Table 1.1. Adaptation of the Haddon matrix in the context of interventions for addressing excess solar UV radiation exposure.

| Phase | Factors | | |
	Human	Intervention	Environment
Pre-event	Genotype and phenotype characteristics (chapter 4) Knowledge, attitudes and behaviour (chapter 5)	Availability of shade, adequate clothing and sunscreen (chapters 6 and 7) Appropriate standards to define the protection expected from physical and chemical agents (chapters 6 and 7)	Terrestrial UV quality and quantity (chapters 2 and 3) Supportive public health policies and guidelines (chapter 5)
Event	Behaviour and time spent outdoors (chapter 5) Use of protective agents (chapters 6 and 7)	Efficacy and appropriate use of sunscreens, clothing, shade and sunglasses (chapters 6 and 7)	Weather—temperature, precipitation, humidity and wind speed (chapters 3 and 5) UV Index (chapters 3 and 5) Desire to be outdoors and to conform with others (chapter 5)
Post-event	Clinical consequences of exposure (chapter 4) Ability to adapt positively to further exposure (chapter 8) Immediately repair the resulting damage (chapter 8)	Development of improved protective agents, such as broad spectrum sunscreens resulting in high compliance (chapters 6 and 7)	Provision of medical and other services capable of treating the damage (chapter 9) On-going monitoring and surveillance (chapter 9)

The matrix is designed so that the linkages and interactions of the various factors before, during and after the injury has incurred can be evaluated leading to a means for identifying and considering possible future resource allocations and activities; relevant research and knowledge, either already available or that needed for the future; and priorities for countermeasures, judged in terms of their costs and their effectiveness in reducing or avoiding undesirable injury or disease. A classic example in the context of what follows are appropriate strategies to counter the year-on-year rise in skin cancer.

Reference

[1] Haddon W Jr 1980 Advances in the epidemiology of injuries as a basis for public policy *Public Health Rep.* **95** 411–21

IOP Publishing

Sun Protection
A risk management approach
Brian Diffey

Chapter 2

The origin and beneficial effects of solar UV radiation

2.1 The nature of optical radiation

In 1666, Isaac Newton '*…procured me a Triangular glass-Prisme, to try therewith the celebrated Phaenomena of Colours*', and opened up a new era into the scientific investigation of light. It was not until 1801 that Johann Ritter discovered the ultraviolet (UV) region of the solar spectrum by showing that chemical action was caused by some form of energy in the dark portion beyond the violet. In the previous year, 1800, Sir William Herschel had demonstrated the existence of radiation beyond the red end of the visible spectrum, a component now known as infrared (IR) radiation.

These three components of the solar spectrum—ultraviolet, visible and infrared—are referred to collectively as *optical radiation*, which is part of the electromagnetic spectrum. Other regions of this spectrum include radio-waves, microwaves, x-rays and gamma radiation. The feature that characterises the properties of any particular region of the spectrum is the wavelength of the radiation.

Optical radiation is divided into three distinct spectral regions: ultraviolet radiation (100–400 nm), visible light (400–780 nm) and infrared radiation (780 nm–1 mm).

Within the UV part of the spectrum the biological effectiveness of the radiation varies strongly with wavelength (see chapter 4) and for this reason the ultraviolet spectrum is subdivided into three regions: UVA, UVB and UVC.

The notion to divide the UV spectrum into different wavebands was first put forward at the Second International Congress on Light held in Copenhagen during August 1932. It was recommended that three spectral regions be defined as follows:

UVA: 400–315 nm
UVB: 315–280 nm
UVC: 280–100 nm

doi:10.1088/978-0-7503-1377-3ch2

The subdivisions are arbitrary and differ somewhat depending on the discipline involved. Environmental and dermatological photobiologists normally define the wavelength regions as:

UVA: 400–320 nm
UVB: 320–290 nm
UVC: 290–200 nm

The division between UVB and UVC is chosen as 290 nm since UV radiation at shorter wavelengths is only present in terrestrial sunlight at high altitude. The choice of 320 nm as the division between UVB and UVA is perhaps more arbitrary. Although radiation at wavelengths shorter than 320 nm is generally more photobiologically active than longer wavelength UV, advances in molecular photobiology indicate that a subdivision at 330–340 nm may be more appropriate and for this reason the UVA region has been divided into UVAI (340–400 nm) and UVAII (320–340 nm).

Like UV radiation, infrared radiation is divided into three wavebands: IRA (780–1400 nm), IRB (1400–3000 nm) and IRC (3000 nm–1 mm).

2.2 The Sun

The Sun is responsible for the development and continued existence of life on Earth. The Sun's infrared rays warm us and we can see with eyes that respond to the visible part of the Sun's spectrum. More importantly, visible light is essential for photosynthesis, the process whereby plants, necessary for our nutrition, derive their energy. Besides serving as the ultimate source of our food and our energy, sunlight also acts on us to alter our chemical composition, control the rate of our maturation and drive our biological rhythms.

The Sun is a very nearly spherical star of diameter 1.39×10^6 km and mean distance from the Earth of 1.5×10^8 km (about 93 million miles). Because it is composed primarily of hydrogen and helium, its density is only 1.4 g cm^{-3}.

The average intensity of solar radiation received by the Earth's atmosphere when the Earth is at its mean distance from the Sun is called the solar constant and is equal to about 1.366 kW m^{-2}.

The solar output is not constant but varies with a 27 day apparent solar rotation, the 11 year cycle of sunspot activity, and occasional solar flares. Because of the elliptical orbit of the Sun, the Sun–Earth distance varies by about 3.4% from a minimum on the perihelion (3 January) to a maximum on the aphelion (5 July). This results in a variation in intensity of about 7% and gives rise to slightly higher radiation levels in southern hemisphere summers than those in the northern hemisphere.

Although the entire Sun is gaseous, the *photosphere* is the part that we see most clearly from Earth and the energy emitted from this region corresponds approximately to a black body temperature of about 5800 K. Atomic absorption and emission occurs in the photosphere and the outer layers of the Sun (the chromosphere and the corona) and produces the characteristic Fraunhofer lines in the

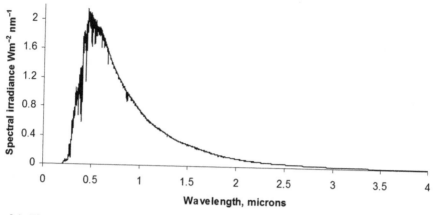

Figure 2.1. The spectrum of extraterrestrial solar radiation (ASTM E-490 Solar Spectra: Air Mass Zero. Available at http://rredc.nrel.gov/solar/spectra/am0/).

visible and UV wavelengths. We see this in figure 2.1, which shows the spectral irradiance of extraterrestrial solar radiation at the mean Sun–Earth distance.

The ultraviolet region accounts for almost 8% of the total solar energy output shown in figure 2.1, but this is reduced to about 5% of the solar radiation that reaches the Earth's surface due to modifying factors in the atmosphere, which are discussed in the next chapter.

2.3 The positive effects of solar UV radiation exposure

Whilst the visible and infrared components of sunlight are essential for fundamental requirements such as nutrition and warmth, is there any need for the UV component, which we now know to be a major risk factor in the causation of skin cancer—the most common human cancer?

Although solar UV radiation is responsible for deleterious effects, notably in skin (see chapter 4), it also has an important role to play in maintaining health and treating disease. The conventional wisdom that the Sun cures human disease originated in ancient times and, to some extent, persists to this day. Many people believe that sunlight is beneficial to health and take the opportunity to expose themselves to it whenever they can. The word *radiation* stems from Aton Ra, the Egyptian sun god, and from Helios, the Greek god of light, is derived the word *heliotherapy*, which means healing by exposure to the Sun's rays.

The best known favourable effect of solar UV radiation in humans (and also other animals) is the production of vitamin D in the skin, but other positive effects include the treatment of some common skin diseases and the sterilisation of water. These benefits are discussed in the remainder of the chapter.

2.3.1 Vitamin D and bone health

The bone disease rickets was reported to occur in humans as early as the second century AD, but it was not considered a significant health problem until people began to congregate in the cities of Northern Europe during the Renaissance. In the mid-17th century, Dr Daniel Whistler was the first to describe this as a disease of English children. He noted that it was associated with deformities of the skeleton, particularly enlargement of the epiphyses at the joints of the long bones and the rib cage, enlargement of the head, bending of the spine, curvature of the thighs and flabby and toneless legs that were usually unable to sustain the weight of the body (see figure 2.2).

The incidence of this debilitating bone disease increased dramatically during the industrial revolution and by the latter part of the 19th century, autopsy studies suggested that approximately 90% of the children raised in the crowded cities had the disease.

However, little attention was focused on the environment as a cause for this disease until 1889 when an investigative committee of the British Medical Association reported that rickets was unknown in the rural districts of Britain, but that it was prevalent in the large industrialised towns. A year later in a classic epidemiological study, Dr Theobald Palm collated clinical observations from a number of physicians throughout the British Empire and reported that, in Great Britain one of the wealthiest nations in the world, rickets abounded, whereas in India and the Orient, where the people were poorest and lived together in squalor, the disease was rare. He concluded that the one factor common to the districts where rickets was unknown was the abundance of sunlight.

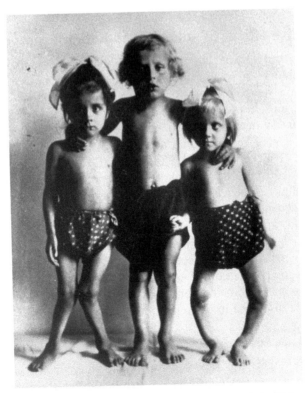

Figure 2.2. Three children with rickets in Vienna circa 1920–30 (courtesy of the Wellcome Library, London, Image ref. L0014375).

Not until the early part of the 20th century was it found that rickets was due to a deficiency of vitamin D and could be treated by exposure to UV radiation, either from sunlight or artificial sources of UV. In fact, the only well-established beneficial effect of solar UV on healthy skin remains the production of vitamin D_3 required for skeletal health. The skin absorbs UVB radiation in sunlight to convert sterol precursors in the skin, such as 7-dehydrocholesterol, to pre-vitamin D_3, followed by heat-isomerisation to vitamin D_3 (cholecalciferol) that is released into the circulation.

Vitamin D exists in two forms: vitamin D_3 produced by the action of sunlight and also contained in some foods such as oily fish e.g. salmon and mackerel, milk and egg yolks; and vitamin D_2 (ergocalciferol) found in some plants.

Irrespective of whether vitamin D is produced in the skin or ingested, it is transported in the blood to the liver where it is converted to 25-hydroxyvitamin D. This molecule is further transformed in the kidneys to form 1,25-hydroxyvitamin D, the biologically active form of vitamin D that is responsible for increasing the efficiency of intestinal calcium absorption. In most people, around 90% of their vitamin D needs results from sun exposure, with diet contributing the remaining 10% or so.

More recently, there is some suggestion that increased production of 1,25-hydroxyvitamin D may protect against colon, breast and prostate cancer and that exposure to UVB with subsequent effects on vitamin D levels may play a role in protecting against other diseases including acute myocardial infarction and multiple sclerosis [1].

So how much sun exposure do we need to maintain healthy vitamin D levels? Adequate sun exposure is not easily defined, but it is generally accepted that exposing the face, hands and arms 2 or 3 times a week to a third to a half of the exposure necessary to result in a just perceptible reddening of white skin in the spring, summer and autumn is more than adequate to satisfy the body's requirement for vitamin D throughout the year.

This exposure can be achieved in 10 min or so lying in the mid-summer sun under a clear sky when some areas of skin are uncovered (figure 2.3a) or, alternatively,

Figure 2.3. Typical outdoor behaviour (a) in a largely unshaded environment and (b) in an urban environment.

exposure for around one hour in an urban environment when normally only the hands and face are exposed and there is partial shading of the sky and sun (figure 2.3b).

At earlier and later times of day and periods outside the summer months, these exposure times increase, as they will also under a cloudy sky. In fact, solar UVB levels are too low in the winter at temperate latitudes to synthesise clinically-significant amounts of vitamin D. However, in healthy individuals, wintertime vitamin D levels are maintained at about two thirds of summertime levels as vitamin D synthesised during the summer is stored in body tissues and released during the winter.

2.3.2 Heliotherapy

Heliotherapy is the treatment of disease by exposure to sunlight. In central Europe around the turn of the 20th century there were several enthusiastic proponents of sunlight and fresh air therapy for a variety of conditions. In his book *Light Treatment in Surgery* published in 1926, Dr Oscar Bernhard writes that for more than 20 years he used sunlight treatment successfully for conditions which included, wounds, osteomyelitis, fractures, syphilitic ulcers, carcinoma of the skin and surgical tuberculosis. General sun baths were taken in the open or in wind protected balconies with a southern aspect, and many of these were established in several European countries, as illustrated in figure 2.4.

Perhaps the most important milestone in the use of sunlight in treating skin disease occurred with the Danish physician Neils Finsen, who was awarded the Nobel Prize in 1903 for his successful treatment of lupus vulgaris, which is a painful cutaneous tuberculosis with a nodular appearance, most often on the face around the nose, eyelids, lips, cheeks, ears and neck. The photograph in figure 2.5 was taken

Figure 2.4. Patients lying outside in the sun on a terrace at Alton Hospital, Hampshire, in 1937 as part of their therapy (courtesy of the Wellcome Library, London, Image ref. L0074520).

Figure 2.5. Treatment of lupus vulgaris with sunlight in Copenhagen c 1895 (courtesy of the Finsen Institute, Copenhagen, Denmark).

in about 1895 in Copenhagen and shows patients with lupus vulgaris under Finsen's care being treated with sunlight.

Although medical opinion of the value of sunlight in treating disease has waned over the last century, it is generally accepted by most skin specialists that many common skin diseases, such as psoriasis, acne and eczema, improve from exposure to summer sunlight. One form of natural phototherapy for psoriasis is that experienced by patients who bask in the sun by the side of the Dead Sea in Israel.

Considerable relief has been reported by many patients, although not all, since in a proportion of patients sunlight can exacerbate the condition. However, dermatologists in some countries prefer to send patients to the Dead Sea (and also the Canary Islands) for four week periods rather than admit them to hospital. Patients normally stay at hotels that incorporate treatment facilities and resident medical staff, as illustrated in figure 2.6.

The mechanism of the treatment is poorly understood. The Dead Sea is 400 m below sea level and sunlight on the shores of the Dead Sea contains much less UVB radiation than at other sites in Israel so that patients can lie in the sun for longer without burning. Also, the Dead Sea has a rich abundance of minerals and organic substances, as well as natural black mud and sulphur springs, amongst which may be an efficient photosensitising material.

2.3.3 Pathogenic benefits of solar UV radiation

In their pioneering studies in 1877, Downes and Blunt showed that sunlight kills bacteria and soon afterwards it was found that it is the UV component that is responsible.

Figure 2.6. The sun-exposure facilities for heliotherapy at the DMZ Medical Spa located on the roof of the Lot Hotel, Ein Bokek, on the shores of the Dead Sea (courtesy of Dr Marco Harari).

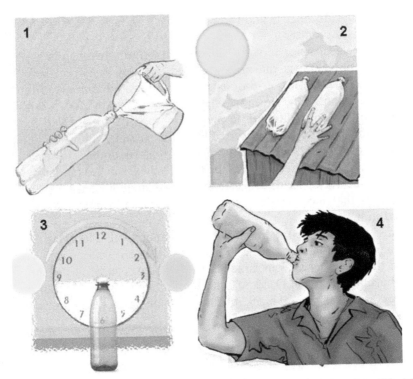

Figure 2.7. The principles of the solar disinfection (SODIS) water treatment technique (reproduced from [2], copyright 2012, with permission from Elsevier).

It is estimated that over 600 million people lack access to suitable drinking water and, along with chlorination and filtration, an inexpensive water treatment that has gained popularity in recent years is that of solar disinfection (SODIS) [2].

The basic SODIS technique is demonstrated in figure 2.7. Transparent containers—normally 2 litre plastic beverage bottles—are filled with contaminated water and placed in direct sunlight for at least 6 h, after which time it is safe to drink.

The germicidal effect of the technique is based on the combined result of thermal heating and UV radiation. It has been shown to be effective for eliminating microbial pathogens and reducing diarrhoeal morbidity, including cholera. Since the technique is cheap and simple to use, the method has spread throughout the developing world and is in daily use in more than 50 countries in Asia, Latin America and Africa, and now more than 5 million people disinfect their drinking water using solar UV radiation [2].

References

[1] Lucas R M, Norval M, Neale R E, Young A R, de Gruijl F R, Takizawag Y and van der Leun J C 2015 The consequences for human health of stratospheric ozone depletion in association with other environmental factors *Photochem. Photobiol. Sci.* **14** 53–87

[2] McGuigan K G, Conroy R M, Mosler H J, du Preez M, Ubomba-Jaswa E and Fernández-Ibáñez P 2012 Solar water disinfection (SODIS): A review from bench-top to roof-top *J. Hazardous Mater.* **235–236** 29–46

Sun Protection
A risk management approach
Brian Diffey

Chapter 3

Factors that influence the quality and quantity of terrestrial solar UV radiation

3.1 The atmosphere

Before reaching the surface of the Earth, solar radiation must pass through the atmosphere, a mixture of gases, which are (by volume): 78.09% nitrogen, 20.95% oxygen, 0.93% argon, 0.04% carbon dioxide, and small amounts of trace gases including helium, neon and methane. The atmosphere also contains a variable amount of water vapour that is typically around 1% at sea level and 0.4% averaged over the entire atmosphere. The process of traversing the atmosphere modifies both the quality (spectrum) and quantity (irradiance) of extraterrestrial solar radiation.

The Earth's atmosphere is divided into five main layers, which are:

Exosphere	700 to 10 000 km above sea level (asl)
Thermosphere	80 to 700 km asl
Mesosphere	50 to 80 km asl
Stratosphere	12 to 50 km asl
Troposphere	0 to 12 km asl

Within these principal layers that are largely determined by temperature, there are a number of secondary layers of which the most important in the context of this book is the ozone layer. The ozone layer is contained within the stratosphere at concentrations of about 2 to 10 parts per million (ppm), with a concentration maximum at an altitude of about 25 km, and is produced by UV radiation at wavelengths less than 242 nm, which breaks the bonds of molecular oxygen (O_2) into atomic oxygen (O). One of the oxygen atoms then combines with molecular oxygen to form ozone (O_3).

The amount of ozone in the atmosphere is balanced by natural processes that destroy ozone due to dissociation of O_3 by UV radiation in the presence of free

doi:10.1088/978-0-7503-1377-3ch3   3-1

radical catalysts, including nitric oxide (NO), nitrous oxide (N_2O) and hydroxyl (OH). With the release of man-made, ozone-destroying substances, such as chloro-fluorocarbons, attention has focused in recent decades on concern that this natural balance is being disturbed.

Although the total ozone column varies seasonally and geographically, it is typically equivalent to a layer of gas about 3 mm thick at normal temperature and pressure at sea level. A total ozone amount of, say, 3.2 mm corresponds to the more commonly used 320 Dobson Units (DU) and represents average conditions in mid-latitudes. Generally the ozone amount is lowest around the equator at approximately 250 DU and increases with latitude to around 420 DU at the polar regions.

The most important property of stratospheric ozone is the absorption of UVC radiation and most of the shortwave UVB radiation that otherwise would reach the Earth's surface and result in damage not only to human skin and eyes but also to plants, animals and marine life.

3.1.1 Solar radiation transport through the atmosphere under clear skies

On its passage through the atmosphere, solar radiation is attenuated principally by the following effects:

- Absorption by atmospheric ozone, which is mainly important for wavelengths less than 330 nm, where the ozone absorption cross section increases rapidly with decreasing wavelength, so that there is practically no radiation with wavelengths less than 290 nm that reach the Earth's surface. This is illustrated in figure 3.1 which compares the spectrum of extraterrestrial solar UV radiation and midsummer terrestrial UV around midday at mid-latitudes on an unshaded, horizontal surface under a cloudless sky at sea level. Also shown in figure 3.1 is the absorption cross section for ozone.
- Rayleigh scattering caused by oxygen, nitrogen and other molecular components of the atmosphere, where the scattering particle is small compared with the wavelength of the radiation. The probability of Rayleigh scattering is inversely proportional to the fourth power of wavelength and as a consequence shorter (blue) wavelengths of light scatter more readily than longer (red) wavelengths. This is why the sky looks blue and also why sunsets are red. At sunset, the Sun is close to the horizon and so the Sun's rays pass through more atmosphere than normal to reach our eyes. Much of the blue light has been scattered away due to the longer pathlength through the atmosphere, leaving mainly red light in a sunset.
- Mie scattering caused by dust, aerosols, and other particles of diameter comparable to the wavelength of the radiation.

Scattered solar radiation is referred to as the *diffuse* component of sunlight (or sometimes *skylight*) and together with the *direct*, or unattenuated, radiation makes up the total, or global, terrestrial solar radiation. Figure 3.2 shows the total, direct and diffuse (i.e. scattered) terrestrial spectral irradiance for a latitude of 45 °N at solar noon in midsummer on an unshaded, horizontal surface under a cloudless

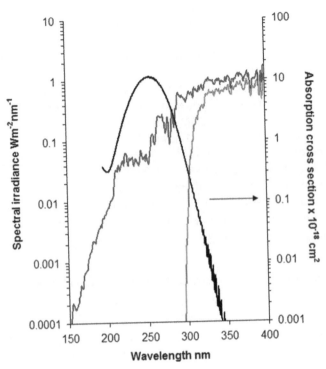

Figure 3.1. The extraterrestrial (blue) spectrum and terrestrial (red) spectral irradiance for a latitude of 45 °N at solar noon in mid-summer. The curve in black is the absorption cross section for ozone.

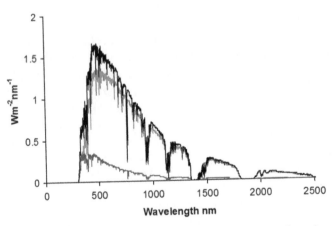

Figure 3.2. The direct (red), diffuse (blue) and global (black) ambient spectral irradiance for a latitude of 45 °N at solar noon in mid-summer under clear skies at sea level.

sky at sea level. Note that the total spectral irradiance peaks at around 500 nm in the blue-green, whereas the diffuse component peaks in the UVA1-blue region of the spectrum, and that infrared radiation comprises almost entirely direct radiation.

3.2 Quantities and units

Quantities of UV radiation are expressed using the radiometric terminology below:

Term	Unit	Symbol
Wavelength	nm	λ
Radiant energy	J	Q
Radiant flux	W	Φ
Radiant intensity	W sr^{-1}	I
Radiance	W m^{-2} sr^{-1}	L
Irradiance	W m^{-2}	E
Radiant exposure	J m^{-2}	H

Terms relating to a beam of radiation passing through space are the *radiant energy* and *radiant flux*. Terms relating to a source of radiation are the *radiation intensity* and the *radiance*. The term *irradiance*, which is the most commonly used term in solar UV studies, relates to the object (e.g. person) struck by the radiation. These radiometric quantities may also be expressed in terms of wavelength by adding the prefix *spectral*. The time integral of the irradiance is strictly termed the *radiant exposure*, but is sometimes expressed as *exposure dose*, or even more loosely as *dose*.

Whilst radiometric terminology is widely used in climatology and photobiology, the units chosen will vary. For example, exposure doses may be quoted in mJ cm^{-2} or kJ m^{-2}. Examples of the equivalence of these units are:

To convert from	To	Multiply by
J cm^{-2}	mJ cm^{-2}	10^3
J cm^{-2}	J m^{-2}	10^4
J m^{-2}	mJ cm^{-2}	10^{-1}
kJ m^{-2}	J cm^{-2}	10^{-1}
kJ m^{-2}	mJ cm^{-2}	10^2

The time required to deliver a UV exposure dose of H J m^{-2} when the irradiance is E W m^{-2} is simply H/E s.

3.2.1 Biologically weighted quantities

As we shall see in chapter 4, the ability of UV radiation to elicit biological responses in human skin depends strongly on wavelength, often encompassing a range of four orders of magnitude between 250 to 400 nm, with the exact wavelength dependence (termed the *action spectrum*—see section 4.1.2) characterised by the biological end-point, such as erythema (sunburn) or skin cancer. Thus a statement that a subject received a certain exposure dose of UV radiation conveys nothing about the consequences of that exposure in terms of biological response.

For example, if a UV exposure of 5 J cm^{-2} was delivered from a UVA lamp, no erythemal response would be seen apart from in people exhibiting severe, abnormal pathological photosensitivity, whereas the same dose delivered from summer sunlight would result in reddening of the skin in most white skinned individuals.

Consequently, there is often a need to express the exposure as a biologically weighted quantity. The biologically weighted irradiance, E_{bio}, is obtained by weighting the spectral irradiance of the source, $E(\lambda)$, with the action spectrum, $\varepsilon(\lambda)$, of the response at each wavelength λ nm and integrating over all wavelengths in the waveband of interest. This is expressed as:

$$E_{bio} = \int_{\lambda} E(\lambda)\, \varepsilon(\lambda)\, \mathrm{d}\lambda$$

If spectral irradiance is expressed in W m^{-2} nm^{-1}, for example, the unit of E_{bio} will be in W m^{-2}. The action spectrum $\varepsilon(\lambda)$ is dimensionless and only expresses the *relative* effectiveness of different wavelengths and so it is necessary to specify its normalisation wavelength; this is normally the wavelength of maximum effect.

Integration of E_{bio} over time (in seconds) results in the biologically weighted dose expressed in J m^{-2}.

Erythema is the most commonly observed response to solar UV radiation in human skin and when $\varepsilon(\lambda)$ refers to the action spectrum for erythema, as defined by the expressions in section 4.2.6, we determine erythemally-effective irradiance (or dose).

The term *standard erythema dose* (*SED*) is often used as a measure of erythemal radiation, and is defined as an erythemally-effective dose of 100 J m^{-2} [1].

3.2.2 The solar ultraviolet index

The solar ultraviolet index (UVI) is a well-established vehicle to raise community awareness and provide information to the public about the potential detrimental effects on health from solar UV exposure. In recent years, the UVI has appeared on TV weather forecasts and internet websites in many countries to provide the public with a guide to the intensity of the Sun's UV rays.

The UVI is a measure of erythemal UV at the Earth's surface and is expressed numerically as the equivalent of multiplying the erythemal-effective irradiance (in W m^{-2}) by 40. So a UVI of 1 is equivalent to an erythemal irradiance of 0.025 W m^{-2} or alternatively, 0.9 SED h^{-1}.

For southern cities in Australia, such as Melbourne (38 °S) and Sydney (34 °S), the measured UVI is maximal during summer (December–February), with a UVI regularly reaching values of 12. In more northerly Australian cities such as Darwin (12 °S), the maximum UVI can reach 16. In summer in Europe, the UVI typically peaks at values from 5 at high latitudes (~60 °N, e.g. Scandinavia) to around 7 in central regions (~50 °N, e.g. southern UK, Belgium, northern France), and up to 9 or 10 in southern Europe (~40 °N, e.g. southern Spain). In the United States, the maximum UVI ranges from 11 in the southern continental United States to 5 in Alaska, whereas in Canada peak values reach 8 in the southern cities.

3.3 Factors affecting the spectral irradiance of terrestrial UV radiation

3.3.1 Solar elevation

The spectral irradiance of terrestrial UV radiation varies with the elevation of the sun above the horizon, or solar altitude. The complementary angle between the sun and the local vertical is termed the solar zenith angle. The solar altitude depends on the time of day, day of year, and geographical location (latitude). The intensity changes because as the solar altitude decreases, i.e. the sun falls lower in the sky, the photon flux emitted by the sun into a given solid angle is distributed over a larger area on the Earth's surface. If we neglect absorption in the atmosphere, the intensity at a solar zenith of $\theta°$ (solar altitude of $90 - \theta°$) is simply equal to the intensity with the sun directly overhead (solar zenith of $0°$) multiplied by cosine θ.

However we cannot neglect attenuation of solar UV in the atmosphere, particularly by ozone, and this serves to absorb UV in a wavelength-dependent manner and hence change the spectrum relative to extraterrestrial sunlight (see figure 3.1). As the sun falls lower in the sky, the pathlength of solar UV radiation through the atmosphere increases, and as a consequence the intensity of UV reaching the Earth's surface decreases at all wavelengths, particularly those shorter than 320 nm.

The solar elevation not only changes the intensity of terrestrial UV radiation but also that in the visible and infrared regions, as illustrated in figure 3.3.

Around noon on a summer's day, UVB (when taken as 290–320 nm) comprises approximately 6% of terrestrial UV, and UVA (when taken as 320–400 nm) the remaining 94%. But since UVB is much more effective than UVA at causing biological damage (see chapter 4), solar UVB contributes about 85% towards a sunburn reaction, with solar UVA contributing the remaining 15%. These percentages presume that visible radiation plays no role in causing sunburn but as we shall see in section 4.2.6, this is probably not the case.

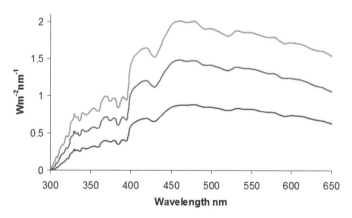

Figure 3.3. Measured global spectral irradiance in Townsville, Australia on 11 January 1996 at solar zenith angles of 5° (red curve), 40° (green curve) and 60° (blue curve) (data plotted from reference [2]).

3.3.2 Surface reflection

Ground reflectance of UV may be important as it can increase ambient levels and result in exposure to parts of the body that might normally be shaded, such as the eyes. Reflection of solar UV radiation from most ground surfaces is normally less than 10%; for example grass reflects about 2%–4% of incident UV and asphalt about 4%–8%. The main exceptions are clean snow that can reflect up to 90% and beach sand, where the reflectance is about 10% upwards depending on the nature of the sand. For example, ooid sand, which is a bright white sand, has a reflectance ranging from 20% at 300 nm to about 40% at 400 nm.

Solar UV reflectance from the ocean
The reflected solar UV radiation from the ocean surface comes from two processes. Some of the direct and diffuse solar radiation incident onto the surface enters the water, there to be scattered one or more times until part of that radiation is re-emitted from the water and part transmitted to deeper depths. The remainder of the incident radiation is reflected upward by the sea surface due to the difference in the refractive indices of air (1.0) and water (approximately 1.34).

The relative contributions of these two effects depend on the constituents of the ocean in terms of phytoplankton, mineral particles, bubbles, etc, on the roughness of the sea surface (the wave state, usually characterised by the wind speed), and on the angular distribution of the incident direct and diffuse UV radiation.

The reflectance of incident solar UV radiation (wavelength range 300–400 nm) for a range of solar altitudes was calculated using representative atmospheric conditions for marine aerosols and marine humidity, and a typical open ocean chlorophyll concentration of 1 mg m^{-3} (figure 3.4). Calculations were carried out for a wind speed of zero, corresponding to a level sea surface, and a fresh breeze of 10 m s^{-1}, which equates to 5 on the Beaufort wind scale and results in a moderate sea state with wave heights of typically 2 m.

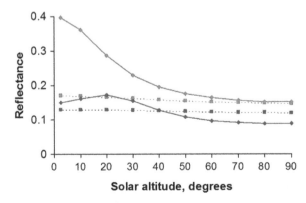

Figure 3.4. Reflectance of UV from the ocean surface as a function of solar altitude for wind speeds of zero (red curves) and 10 m s^{-1} (blue curves) for both clear (solid curves) and overcast (broken curves) sky conditions (values generated by Dr Curtis Mobley using the HydroLight version 5.3 radiative transfer software www.hydrolight.info).

The reflectance plotted in figure 3.4 is the ratio of the total reflected scalar irradiance to the total incident scalar irradiance. Scalar irradiance is computed by summing the energy travelling in all upward or downward directions, with each direction weighted equally. We see from figure 3.4 that reflectance is generally between 10%–20%. When the sky is clear, the reflectance starts to rise at solar altitudes below about 30° because of the rapid increase in Fresnel reflectance of a calm sea surface at low solar altitudes. This is only true for a level sea surface since for a rougher sea less UV is reflected because of the effect of tilted wave facets and the associated Fresnel reflectance effects. The change in reflectance at solar altitudes below and above 30° from a calm sea is illustrated in figure 3.5, which shows the

Figure 3.5. Reflection of sunlight from the ocean surface at a solar altitude of 16° (upper image) and 48° (lower image).

reflection of sunlight from the surface of the sea at Sidmouth (latitude 50.6 °N) at solar noon, i.e. looking due south, on 30 December (upper image) and 1 September (lower image) when the solar altitudes were 16 degrees and 48 degrees, respectively.

In both cases, the sun was not obscured by clouds. It is apparent that much more sunlight is reflected at the lower solar altitude compared with the image taken on 1 September.

When the sky is overcast the reflectance is virtually independent of solar altitude with only a small difference between a calm and a choppy sea. This arises because the radiance distribution of UV from a heavy overcast sky is almost independent of the solar altitude. For a given set of conditions, the spectral reflectance is virtually independent of wavelength over the UV waveband, although the absolute values of spectral reflectance depend upon the boundary conditions i.e. clear/overcast and no wind/fresh breeze.

For a clear sky and fresh breeze (wind speed of 10 m s^{-1}) at low solar altitudes, the incident rays tend to hit the 'front side' of waves that are tilted toward the sun. This effectively decreases the incident angle, reducing the reflectance. The 'back-sides' of the waves are hit less often because they are on the shadow side of the waves. The consequence of this is that wind-blown surfaces let more energy into the ocean than calm surfaces, as is evident from the blue curves in figure 3.4.

This is especially important in the Arctic, where the sun is usually near the horizon since more sunlight getting into the Arctic ocean because of a rough water surface (rather than highly reflective ice) warms the ocean, melts more ice and so sets up a positive feedback loop of global warming.

Clear waters generate less re-emitted radiation than do turbid waters with high concentrations of scattering particles. A calm sea surface and a clear sky give strong specular reflection of the sun's rays—the so-called glitter pattern of sparkles often seen on a water surface. However, the ocean is rarely glassy smooth and the impact of wind-blown surfaces give rise to a shimmering effect, most noticeable at low solar altitudes, as illustrated in the upper image of figure 3.5.

A rough surface or an overcast sky gives more diffuse reflectance of the incident radiation. How much direct or diffuse UV radiation is reflected depends strongly on the solar altitude and the surface roughness.

3.3.3 Aerosols and air pollution

Aerosols are small particles, such as smoke, salt, smog, ash and dust, that are suspended in air and usually occur at altitudes below 2 km. Sizes of aerosol particles vary from around 10 nm up to a few tens of microns. Although aerosols can attenuate solar UV radiation, the effect is small for a low turbidity atmosphere with just a 5% or so reduction in erythemal effective irradiance relative to zero turbidity, increasing to about 25%–35% reduction for a high turbidity atmosphere.

Absorption by pollutant gases such as NO_2 and SO_2 can decrease UVB by a few per cent in heavily polluted urban areas. In exceptional circumstances when the levels of pollutants or dust are very high, solar UV radiation is significantly reduced, for example after the Mount Pinatubo eruption in 1991.

3.3.4 Altitude

Solar UV radiation increases with altitude above sea level at about 5%–8% per km. At high elevations, for example in the Alps, where there is lower pollution levels and often less cloud cover, the increase is often larger.

3.3.5 Clouds

The condensation of water vapour, which occurs mainly in the troposphere, leads to the formation of clouds, which are simply a suspension of water droplets, ice crystals, or both.

Clouds can be classified as layered (*strato-*) or convective (*cumulo-*) and also as low/middle (*alto-*), high (*cirro-*), or precipitating (*nimbo-*). The ten basic cloud types are cirrus, cirrostratus, cirrocumulus, altostratus, altocumulus, cumulus, stratus, stratocumulus, nimbostratus, and cumulonimbus (thunderstorm cloud), and are illustrated in figure 3.6.

Typical values for the frequency of occurrence of cloud types over land and ocean in northern mid-latitudes (30 °N to 60 °N) averaged over the year are shown in table 3.1.

Cloudy conditions dominate the weather at mid-latitudes and are a major factor in modulating solar radiation. Sky conditions are estimated in terms of how many eighths of the sky are covered in cloud, ranging from 0 oktas (completely clear sky) through to 8 oktas (completely overcast). The erythemally-effective cloud transmittance, defined as the ratio of irradiance under cloudy skies to that under clear skies, can be roughly expressed as $1 - A \times c^3$, where c is the fractional degree of cloud cover,

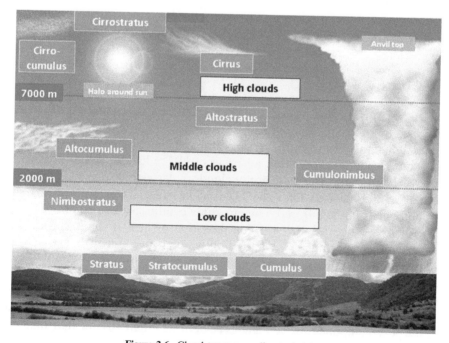

Figure 3.6. Cloud types according to height.

Table 3.1. Average values for frequency of occurrence of different cloud types over land and ocean at mid-latitudes in the northern hemisphere (data taken from reference [3]).

Cloud type	% frequency of occurrence	
	Land	Ocean
Clear sky	23	7
Cirrus, cirrostratus, cirrocumulus	45	32
Altostratus, altocumulus	32	44
Stratus, atratocumulus	28	54
Cumulus	11	24
Cumulonimbus	8	7
Nimbostratus	12	10

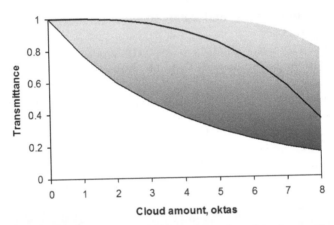

Figure 3.7. The erythemally-effective cloud transmittance as a function of cloud cover. The black curve is an average weighted by the relative frequency of occurrence of different cloud types over land (table 3.1), whilst the shaded area approximates to the range of instantaneous values ranging from a covered (lower bound) to an uncovered (upper bound) sun.

expressed in oktas divided by 8, and A is a coefficient depending on the particular cloud type [4]. The coefficient, A, is 0.40, 0.69, 0.75, 0.88 and 1.06 for cirrus, alto-cumulus/alto-stratus, strato-cumulus, cumulus, and cumulo-nimbus, respectively.

Since pure water is a very weak absorber of UV radiation, clouds attenuate UV primarily by scattering. The influence of clouds on terrestrial solar radiation is more important in the infrared than the UV region, since water vapour has several absorption bands in the infrared. Under cloudy conditions the risk of overexposure may be increased because the warning sensation of heat is diminished. Roughly speaking the ambient annual UV radiation is about two thirds that estimated for clear skies in mid-latitudes, rising to about 75% for the tropics.

Light clouds scattered over a blue sky make little difference to UV intensity unless directly covering the sun, whilst complete light cloud cover reduces terrestrial UV to under one half of that from a clear sky. Even with heavy cloud cover the scattered, or diffuse, UV component of sunlight is seldom less than 10% of that under clear sky. However, very heavy storm clouds can virtually eliminate terrestrial UV even in summertime. The effect of cloud cover on erythemal UV radiation is illustrated in figure 3.7.

3.4 Measuring solar UV radiation

The measurement of ambient solar UV radiation is performed with scanning spectroradiometers, array spectroradiometers, multi-channel filter radiometers, and broadband radiometers.

There are several objectives of solar UV measurement, which include:
- Establishing UV climatology by long-term monitoring.
- Detecting trends in global UV irradiance.
- Providing datasets for the validation of computational models of UV climatology.
- Understanding geographical differences in global UV climatology.
- Providing data for public information and awareness (e.g. UV Index).

3.4.1 Scanning spectroradiometers

The fundamental way to characterise the nature of either natural or artificial solar UV radiation is to measure the spectral power distribution. This is a graph (or table) that indicates the radiative power received on a surface as a function of wavelength. The data are obtained by a technique known as spectroradiometry. The three basic requirements of a spectrometer system are the *input optics*, designed to conduct the radiation from the source into the *monochromator*, which disperses the radiation onto a *detector*.

Input optics

The spectral transmission characteristics of monochromators depend upon the angular distribution and polarisation of the incident radiation as well as the position of the beam on the entrance slit. For measurement of spectral irradiance, direct irradiation of the entrance slit should be avoided. An input optic should be used to ensure that the system has a cosine-weighted response; this means that the incident flux is weighted by the cosine of the angle between the incoming radiation and the normal to the surface. Input optics are generally one of two types: an integrating sphere or a diffuser made of polytetrafluoroethylene (PTFE), also known as Teflon™.

Monochromator

A ruled diffraction grating is normally preferred to a prism as the dispersion element in the monochromator used in a spectroradiometer, mainly because of better stray radiation characteristics. High-performance spectroradiometers, used for determining low UV spectral irradiances in the presence of high irradiances at longer wavelengths, demand extremely low stray radiation levels. Such systems must incorporate a double monochromator, that is, two single ruled grating monochromators in tandem.

It is important that spectroradiometers are calibrated over the wavelength range of interest using standard lamps. A halogen lamp operating at a colour temperature of

Figure 3.8. Input optics of an SUV-100 spectroradiometer (Biospherical Instruments Inc.) at the installation at the South Pole. The temperature-stabilised, scanning, double monochromator coupled to a photomultiplier tube detector is housed inside the building (courtesy of Dr Germar Bernhard, Biospherical Instruments Inc. San Diego, USA).

about 3000 K can be used as a standard lamp for the spectral interval 250–2500 nm, although for measurements concerned solely with the ultraviolet region (200–400 nm), a deuterium lamp may be used.

Accurate spectroradiometry requires careful attention to detail. Factors that can affect accuracy include wavelength calibration, bandwidth, stray radiation, polarisation, angular dependence, the linearity and dynamic range of the photodetector, and variations in temperature, especially when used to measure solar spectral irradiance in harsh environments, as illustrated in figure 3.8.

A particular type of spectroradiometer used in UV climatology measurement is the Brewer instrument. This is an automated, diffraction-grating spectroradiometer that provides near-simultaneous observations of the sun's intensity at six wavelengths in the range 295–365 nm. These data are used to calculate ozone and SO_2 levels, as well as aerosol optical depth.

3.4.2 Array spectroradiometers

One problem with a monochromator is that it often takes several minutes to scan a spectrum, depending on factors such as wavelength interval, bandwidth and time spent sampling the signal at each wavelength. During the time of the scan, spectral irradiance can change due to variable meteorological conditions and changing solar elevation. The alternative is to disperse the different wavelengths of the input spectrum onto an output plane where, for example, a charge coupled detector (CCD) or diode array is used as the detector. The drawback of this device is that the stray light characteristics are not as good as with a double monochromator, a serious shortcoming when measuring solar UV radiation, especially at wavelengths less than 320 nm where spectral irradiance changes rapidly with wavelength due to the rapid rise in the ozone absorption coefficient.

3.4.3 Multi-channel filter radiometers

Multi-channel filter radiometers (MCFRs) are designed to make measurements in several discrete wavelength bands. The wavelength selection is typically achieved by narrow-to-moderate band interference filters, and the signal is detected with a photodiode or a phototube. Typically, the number of channels is larger than two, and some examples of these instruments include shadow bands which enable the near-simultaneous measurement of diffuse and direct irradiances, in addition to global irradiance. These instruments can be used to reconstruct spectra of solar global irradiance, to derive specific measures such as erythemally-weighted irradiance, or to determine total column ozone.

3.4.4 Broadband radiometry

Although spectroradiometry is the fundamental way to characterise radiant emission, the high cost of spectroradiometers means that solar UV radiation levels are more frequently measured by techniques of broadband radiometry. Broadband radiometers generally combine a detector, e.g. vacuum phototube or a solid-state photodiode, with a wavelength-selective device, e.g. colour glass filter or interference filter, and suitable input optics e.g. PTFE diffuser.

By selecting a detector and filter(s) with the appropriate spectral sensitivity and transmission, respectively, broadband radiometers can be tailored to respond optimally to radiation in a given waveband, such as the UVA region.

Of particular photobiological interest is to use a broadband radiometer whose spectral response matches the erythema action spectrum as closely as possible. This can be achieved reasonably well by appropriate choice of detector/filter combination. The earliest instrument that had this property was the Robertson–Berger meter but now improved models are available. The result is that the radiometer can be calibrated to read the erythemally-weighted irradiance directly. The reading is then correct for any spectral power distribution with an accuracy limited by the degree to which the radiometer's spectral response differs from the erythema action spectrum.

Confidence in the uncertainty of measured solar UV irradiance is important if data from one site are to be compared with other sites. As a result, international inter-comparisons of solar UV monitoring systems are carried out from time to time. Figure 3.9 shows 36 UV broadband radiometers from 16 countries being calibrated in the Swiss Alps at 1610 m asl in 2006. Various types of radiometer were represented in the campaign, but all of them were designed to measure erythemally-weighted irradiance.

Much more detailed information on these spectral and broadband instruments can be found in the series of publications by the World Meteorological Organization [5–8].

Figure 3.9. Inter-comparison of 36 broadband radiometers on a roof platform in the Swiss Alps (courtesy of Dr Gregor Hülsen, Physical Meteorological Observatory, Davos).

3.4.5 Personal UV dosimeters

Whilst radiometers and spectroradiometers are used to monitor ambient solar UV radiation, these instruments are not suited to determining the UV exposure of individuals, especially at multiple sites over the body, because of their bulk and generally high cost. Ideally UV dosimeters designed for personal use should have the following characteristics:

- The dosimeter should be easy to handle and not impose restrictions on the activities of the wearers.
- In physical dosimeters, the output electrical signal should increase linearly with irradiance and have a wide dynamic range.
- In chemical or biological dosimeters the change produced in the system should increase linearly with UV dose. If not, the dose response curve should at least be monotonic, that is, any given dosimeter response is the result of only one radiation dose.
- The dosimeter should exhibit photoaddition; in other words, each wavelength acts independently and the effect of polychromatic radiation is the sum of the effects of all wavelengths involved.
- The spectral sensitivity of the dosimeter should, ideally, match the action spectrum of the photobiological effect being monitored, most commonly erythema.
- The dosimeter response should be independent of temperature and humidity, it should exhibit no 'dark effect' (continuing response when radiation exposure terminated), and it should be stable on long-term storage.

Figure 3.10. An electronic time-stamped UV dosimeter worn on the wrist (courtesy of Dr Elisabeth Thieden and MS(Engin) Jakob Heydenreich).

- The dosimeter should not require laborious processing and should be easy to convert the physical, chemical or biological response to a measure of UV exposure dose.
- The cost per dosimeter should be low so that large scale monitoring is feasible.

Physical dosimeters

The availability of miniature electro-optical UV sensors means that it is possible to construct small UV detectors that can be electrically coupled to a data logger clipped to clothing or worn on the wrist. By this means it is possible to sample UV irradiance in short time intervals over many hours. An example of this type of device, developed by scientists at Bispebjerg Hospital in Copenhagen [9], is shown in figure 3.10.

Chemical dosimeters

The use of chemical methods, which measure the chemical change produced by UV radiation, is called *actinometry*. The most commonly used material for studies of personal UV dosimetry has been the thermoplastic, polysulphone [10]. The basis of the method is that when film is exposed to UV radiation at wavelengths principally in the UVB waveband, its UV absorption increases. This increase, measured at 330 nm before and after exposure in a conventional UV spectrophotometer, gives a measure of the erythemal dose received by the film badge. Normally the film (30–50 μm thick) is mounted in cardboard or plastic holders and worn on a lapel site (see figure 3.11).

Applications of personal UV dosimetry using polysulphone film have included sun exposure of children and adults, sun exposure from different leisure pursuits and occupations, anatomical distribution of sunlight in manikins, humans and animals,

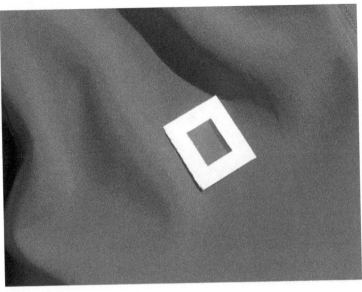

Figure 3.11. A polysulphone UV-sensitive film badge.

as well as UV exposure of patients from therapeutic light sources and UV exposure of workers from industrial UV sources [10].

Biological dosimeters
Biological techniques of measurement are generally limited to the use of viruses and micro-organisms. The cost of biological dosimeters is greater than for polysulphone film and this coupled with a more tedious processing procedure, means they have not found widespread use in studies of personal exposure to sunlight.

3.5 Ultraviolet climatology

Several laboratories around the world continuously monitor erythemal, and sometimes UVA, ambient radiation on an unshaded, horizontal surface giving an unimpeded view of the whole sky, and many of these data are readily available via the internet. For example, The Australian Radiation Protection and Nuclear Safety Agency (ARPANSA) has a network of UV detectors in cities around Australia and provides real-time data on UV levels (see www.arpansa.gov.au/uvindex/realtime/).

On a clear day, the irradiance of both ambient erythemal UV and UVA radiation follows an approximate Gaussian curve (figure 3.12). The width of the distribution depends upon daylength, which itself is a function of latitude and time of year, but for a given daylength the temporal variation of UVA always shows a greater width than for erythemal UV.

Most days, especially at mid-latitudes, are not cloudless and the variation of daily UV exposure throughout the year exhibits perturbations largely due to day-to-day changes in weather, as exemplified in figure 3.13.

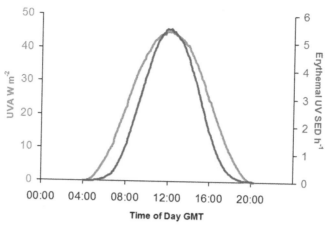

Figure 3.12. The variation of ambient erythemal UV (blue curve) and UVA (red curve) irradiance measured during a clear summer day in Durham, UK (latitude 54.8 °N).

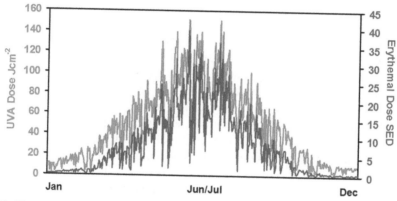

Figure 3.13. The diurnal variation of ambient erythemal UV (blue curve) and UVA (red curve) radiation during 1994 measured in Durham, UK (latitude 54.8 °N).

Note from figures 3.12 and 3.13 that both UVA and erythemal UV (largely UVB) exhibit temporal variations over the course of a day and also a year and yet it is not uncommon to read in some cosmetic and dermatological literature phrases such as: *'It is useful to remember that the level of UVA reaching the Earth's surface is very similar in both summer and winter'* and *'UVB intensity declines from noontime apex, but UVA intensity remains relatively constant throughout the day'*. Misinformation such as this may have contributed to the promotion of the need for year-round sun protection even at northerly latitudes.

Measurements of ambient annual erythemal UV from a number of ground-based networks around the world are shown in figure 3.14, which include data from both the northern and southern hemispheres.

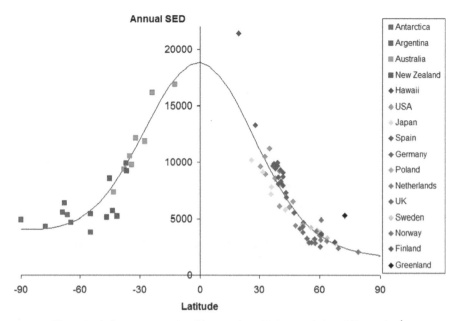

Figure 3.14. Average annual ambient erythemal UV recorded at different sites[1].

Note that although there is a good deal of scatter around the smooth curve, due to factors such as local weather patterns, pollution, differences in altitude above sea level, instrument calibration accuracy, and the period over which annual data have been averaged, the latitude gradient is clearly evident with increasing annual erythemal UV as locations move closer to the equator.

The effect of altitude, in particular, is seen clearly in the data from Hawaii (Mauna Loa Observatory; dark red triangle) and Greenland (Summit Camp Research Station; black triangle), where both measurement sites are over 3000 m above sea level (asl). Also, the South Pole is 2835 m asl, and so its high altitude, combined with the 'ozone hole', will result in its relatively high annual ambient UV radiation.

UV levels in the southern hemisphere tend to be higher than corresponding latitudes in the northern hemisphere due to a cleaner atmosphere, a thinner ozone layer (especially in the Antarctic), and the impact of the elliptical orbit of the Earth around the Sun. Furthermore, broken clouds at sites in the polar regions can enhance UV irradiance above clear sky values due to multiple reflections between the snow-covered surface and the cloud ceiling.

---

[1] I am grateful to the following people who kindly provided one or more of the data points in figure 3.14: Dr Germar Bernhard, Biospherical Instruments Inc., San Diego, USA; Dr Peter Gies, Australian Radiation Protection and Nuclear Safety Agency; Anu Heikkilä and Kaisa Lakkala, Finnish Meteorological Institute; Becky Hooke, Public Health England; Dr Bjørn Johnsen, Norwegian Radiation Protection Authority; Dr Weine Josefsson, Swedish Meteorological and Hydrological Institute; Dr Richard McKenzie, The National Institute of Water and Atmospheric Research, Lauder, New Zealand. In addition some data were extracted from [16–18].

3.6 Reference solar UV spectrum

It is clear that the spectrum of sunlight is continuously changing as the sun rises and falls. For many purposes, however, it is necessary to use some reference solar UV spectrum, and two spectra that are often employed in such calculations are those measured on a clear summer's day around noon at Melbourne (38 °S) and Albuquerque (38 °N). Numerical values of the spectral irradiance at these two sites from 290 to 400 nm in 1 nm steps are given in table 3.2.

Although both locations are at the same distance from the equator and measurements were made close to the respective summer solstices, note that spectral irradiance values were higher in the southern hemisphere compared with the northern hemisphere. Two factors will contribute to this difference; the Earth's elliptical orbit means it is closest to the sun on 3 January and furthest away on 4 July, and that there is generally a clearer atmosphere in the southern hemisphere.

The spectral irradiance measured in Albuquerque has been adopted by the trade organisation *Cosmetics Europe* as a 'standard sun' UV spectrum [11] and is commonly used by the cosmetics industry for calculating the protection provided by topical sunscreens.

3.7 Simulated sources of sunlight

In many photochemical and photobiological studies it is valuable to be able to simulate natural sunlight in the laboratory. The most common way to achieve this endpoint is to use an optically filtered xenon arc lamp as a so-called *solar simulator*.

The xenon lamp has a continuous spectrum throughout the UV, visible and near infrared regions and by appropriate selection of optical filters, a close spectral match can be obtained with natural sunlight, as illustrated in figure 3.15.

Solar simulators vary in power up to 1 kW and higher, and offer uniform beam areas up to 30×30 cm.

3.8 Modelling solar UV radiation

Measurement of ambient UV radiation demands specialist equipment and careful attention to calibration and stability. Consequently, models to compute solar UV

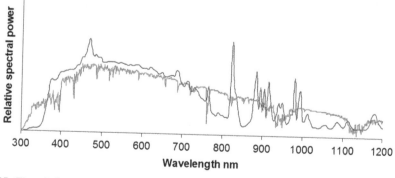

Figure 3.15. The relative spectral power of clear sky, terrestrial UV radiation for mid-latitudes at noon in summer (red curve) and an Oriel® Solar Simulator incorporating an air mass filter AM1.5 (blue curve).

Table 3.2. Solar spectral irradiance in units of W m^2 nm^{-1} at solar noon on clear summer days measured in Albuquerque (38 °N) on 3 July 1986 and in Melbourne (38 °S) on 17 January 1990.

λ, nm	38 °N	38 °S	λ, nm	38 °N	38 °S	λ, nm	38 °N	38 °S
290	0.0000	0.0000	327	0.473	0.618	364	0.648	0.808
291	0.0000	0.0000	328	0.501	0.566	365	0.683	0.766
292	0.0000	0.0001	329	0.517	0.629	366	0.718	0.967
293	0.0001	0.0002	330	0.532	0.708	367	0.740	0.911
294	0.0002	0.0004	331	0.533	0.612	368	0.762	0.861
295	0.0007	0.0009	332	0.533	0.632	369	0.764	0.872
296	0.0012	0.0019	333	0.528	0.630	370	0.766	0.975
297	0.0023	0.0030	334	0.523	0.601	371	0.758	0.856
298	0.0033	0.0048	335	0.514	0.667	372	0.750	0.814
299	0.0060	0.0089	336	0.504	0.575	373	0.706	0.787
300	0.0086	0.0111	337	0.502	0.536	374	0.661	0.705
301	0.0161	0.0196	338	0.499	0.617	375	0.664	0.685
302	0.0236	0.0235	339	0.519	0.660	376	0.666	0.845
303	0.0335	0.0513	340	0.539	0.765	377	0.706	0.876
304	0.0435	0.0574	341	0.549	0.650	378	0.746	1.10
305	0.0577	0.0807	342	0.559	0.680	379	0.750	0.917
306	0.0719	0.0812	343	0.547	0.719	380	0.754	0.839
307	0.0844	0.113	344	0.535	0.570	381	0.698	0.957
308	0.0968	0.135	345	0.535	0.640	382	0.642	0.693
309	0.115	0.127	346	0.534	0.642	383	0.614	0.543
310	0.134	0.147	347	0.536	0.680	384	0.585	0.587
311	0.155	0.235	348	0.537	0.638	385	0.605	0.834
312	0.175	0.215	349	0.548	0.640	386	0.626	0.724
313	0.194	0.246	350	0.559	0.724	387	0.649	0.775
314	0.213	0.269	351	0.574	0.743	388	0.672	0.765
315	0.228	0.283	352	0.589	0.717	389	0.715	0.795
316	0.243	0.243	353	0.601	0.695	390	0.757	0.948
317	0.261	0.371	354	0.613	0.829	391	0.737	1.03
318	0.279	0.316	355	0.608	0.832	392	0.716	0.948
319	0.297	0.353	356	0.603	0.757	393	0.686	0.494
320	0.314	0.401	357	0.570	0.603	394	0.655	0.609
321	0.323	0.400	358	0.538	0.582	395	0.668	0.988
322	0.332	0.405	359	0.551	0.594	396	0.681	0.862
323	0.346	0.359	360	0.564	0.854	397	0.741	0.510
324	0.361	0.444	361	0.582	0.669	398	0.801	1.02
325	0.403	0.448	362	0.600	0.671	399	0.906	1.25
326	0.445	0.600	363	0.624	0.795	400	1.01	1.27

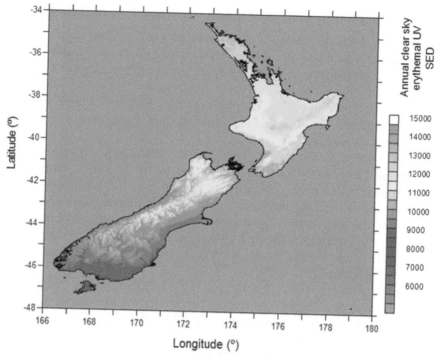

Figure 3.16. A map of clear sky annual erythemal doses over New Zealand (reprinted with permission from [13]).

irradiance are a useful tool for studying the impact of variables on the UV climate.

There are a number of computational models of solar spectral irradiance ranging from radiative transfer models, e.g. [12], to empirically based models. Many of these models have been developed by atmospheric scientists and tend to be complex and require many different inputs that can include type of environment, humidity, ground reflectance, altitude, turbidity, total precipitable water, and ozone column. The equations describing the models can be formidable and their implementation presents a challenge for many biological and clinical scientists, as does knowledge on what values should be entered for many of the atmospheric parameters.

However, user-friendly internet-based interfaces are available such that non-specialists can benefit from complex radiative transfer models without the need to understand the detail. One such example is shown in figure 3.16, which shows the output produced from modelling clear sky solar UV radiation using a radiative transfer model, with ozone and surface pressure as inputs.

The package was developed by the National Institute of Water and Atmospheric Research (NIWA), New Zealand [13]. Figure 3.16 illustrates a map of annual clear sky erythemal doses over New Zealand calculated for the year 2001. The map excludes cloud effects and so although the calculations are made for a specific year (2001), the map would look very similar for any year as ozone changes are less than 5% from year to year, and aerosol effects are small in New Zealand.

3.9 Ozone depletion and its impacts on solar UV radiation

In 1995, Mario Molina, Sherwood Rowland and Paul Crutzen were awarded the Nobel Prize for Chemistry for their predictions 20 years earlier that man-made chlorine compounds released at ground level would diffuse into the upper atmosphere and destroy the ozone resident there. Furthermore, in the mid-1980s scientists from the British Antarctic Survey showed that each year since the 1970s there had been an unexpected decrease in the abundance of springtime ozone over the Antarctic—the ozone hole.

As a consequence, for the past 40 years there has been concern that anthropogenic damage to the Earth's stratospheric ozone layer by the release of ozone-depleting substances, such as chlorofluorocarbons, at ground level will lead to an increase of solar UV radiation reaching the Earth's surface, with a consequent adverse impact not just on human health, but also detrimental to animal, fish and plant life. More recently, there has been an increased awareness of the interactions between ozone depletion and climate change (global warming), which could also impact on human exposure to terrestrial UV.

Over the intervening years, the United Nations Environment Programme has published a number of reports assessing current knowledge on trends in ozone depletion and its impact on terrestrial solar UV radiation and on environmental consequences, including human health. The most recent assessment was published in 2015 [14] and some of its findings are summarised in sections 3.9.1 and 3.9.2.

3.9.1 Current status of atmospheric ozone

During the period 2008–12 mean values of ozone relative to the 1964–80 mean values were smaller by ~3.5% in the Northern Hemisphere mid-latitudes (35 °N–60 °N) and by ~6% in the southern hemisphere mid-latitudes (35°S–60°S). In the tropics (20 °S–20 °N), no significant changes have occurred in total ozone over this period.

Following the decline in total ozone between the 1960s and 1990s, the levels of total ozone outside the polar regions have stopped decreasing since the late 1990s, consistent with the slow decline of ozone-depleting substances over the same period. A number of studies now indicate that total ozone has increased by ~1% since 2000 in the latitude band 60 °S–60 °N in response to stratospheric ozone recovery.

By contrast, at high latitudes (63°–90°) of both hemispheres, ozone depletion continues to occur during winter and spring. Compared to the average values before 1980, the 2010–13 mean total ozone is lower by ~27% in the southern hemisphere in October and by ~10% in the northern hemisphere in March. The Antarctic ozone hole has continued to appear each spring.

Because of the success of the Montreal Protocol and its subsequent amendments in reducing the ozone depleting substances, a gradual recovery of ozone is expected in the decades ahead; indeed, by the end of the 21st century, amounts of ozone in most regions are expected to be greater than they were before ozone depletion began prior to 1980.

3.9.2 Trends in ambient UV radiation

There are a number of ground-based networks providing data on ambient UV radiation, some going back for more than 20 years, with instruments deployed mostly in Australia, New Zealand, Europe, the US, Canada, South America, and both polar regions, with a scarcity of data available from Africa, the Middle East, and Asia.

With the recovery of stratospheric ozone to levels before the 1980s, it might be expected that this change is reflected in trends of UV radiation measurements. At most locations, however, any trends are currently below the detection threshold imposed by the limited period of most UV monitoring networks, accuracy of instrument calibration and long-term stability of monitoring equipment, year-to-year fluctuations in cloud cover, and an increase in ozone and aerosols present in the lower atmosphere due to pollution. Nevertheless, international inter-comparisons of instrumentation are improving data quality and reliability.

Overall, it appears that ozone will continue to be the dominant factor affecting ambient UV levels over Antarctica throughout the 21st century, but that the effects of aerosols, although highly uncertain, will probably dominate future changes in ambient UV radiation in highly populated regions.

On the occasion of the 20th anniversary of the Montreal Protocol in 2007, a number of eminent scientists who had gathered in Athens noted that the success of the Montreal Protocol was largely based on scientific progress made over the last two decades [15]. A world of extreme high chlorine, low ozone, and high UV has presently been avoided and continued adherence to the Montreal Protocol will assure that this remains the case in the future. It is perhaps one of the most illustrious examples of a successful global collaboration between scientific, industrial and environmental organisations, and policy makers. We should not be complacent, however, since factors that could influence recovery include non-compliance with the Montreal protocol, interactions between ozone depletion and global warming, and future volcanic eruptions.

References

[1] Commission Internationale de l'Eclairage 1998 *Erythema Reference Action Spectrum and Standard Erythema Dose.* CIE S007E-1998 (Vienna: CIE Central Bureau)

[2] Bernhard G, Mayer B, Seckmeyer G and Moise A 1997 Measurements of spectral solar UV irradiance in tropical Australia *J. Geophys. Res.* **102** 8719–30

[3] Warren S G, Hahn C J and London J 1985 Simultaneous occurrence of different cloud types *J. Clim. Appl. Meterol.* **24** 658–67

[4] Thiel S, Steiner K and Seidlitz H K 1997 Modification of global erythemally effective irradiance by clouds *Photochem. Photobiol.* **65** 969–73

[5] Seckmeyer G, Bais A, Bernhard G, Blumthaler M, Booth C R, Disterhoft P, Eriksen P, McKenzie R L, Miyauchi M and Roy C 2001 Instruments to measure solar ultraviolet irradiance. Part 1: Spectral instruments, Global Atmosphere Watch, Report No. 125 (Geneva: World Meteorological Organization)

[6] Seckmeyer G, Bais A, Bernhard G, Blumthaler M, Booth C R, Lantz K and McKenzie R L 2005 Instruments to measure solar ultraviolet irradiance. Part 2: Broadband instruments

measuring erythemally weighted solar irradiance, Global Atmosphere Watch, Report No. 164 (Geneva: World Meteorological Organization)

[7] Seckmeyer G, Bais A, Bernhard G, Blumthaler M, Johnsen B, Lantz K and McKenzie R L 2010 Instruments to measure solar ultraviolet irradiance. Part 3: Multi-channel filter instruments, Global Atmosphere Watch, Report No. 190 (Geneva: World Meteorological Organization)

[8] Seckmeyer G, Bais A, Bernhard G, Blumthaler M, Drüke S, Kiedron P, Lantz K, McKenzie R L and Riechelmann S 2010 Instruments to measure solar ultraviolet radiation. Part 4: Array spectroradiometers, Global Atmosphere Watch, Report No. 191 (Geneva: World Meteorological Organization)

[9] Heydenreich J and Wulf H C 2005 Miniature personal electronic UVR dosimeter with erythemal response and time-stamped readings in a wristwatch *Photochem. Photobiol.* **81** 1138–44

[10] Diffey B L 1989 Ultraviolet radiation dosimetry with polysulphone film *Radiation Measurement in Photobiology* ed B L Diffey (London: Academic) 135–9

[11] COLIPA 1994 *SPF Test Method* (Brussels: European Cosmetic Toiletry, and Perfumery Association)

[12] Mayer B and Kylling A 2005 The libRadtran software package for radiative transfer calculations—description and examples of use *Atmos. Chem. Phys.* **5** 1855–77

[13] Bodeker G E, Shiona H, Scott-Weekly R, Oltmanns K, King P, Chisholm H and McKenzie R L 2006 UV Atlas version 2: What you get for your money, paper presented at *UV Radiation and its Effects: an update (Dunedin, NZ, 19–21 April 2006)* Available at www.niwa.co.nz/sites/niwa.co.nz/files/import/attachments/Bodeker.pdf (last accessed 17 March 2017)

[14] Bais A F, McKenzie R L, Bernhard G, Aucamp P J, Ilyas M, Madronich S and Tourpali K 2015 Ozone depletion and climate change: impacts on UV radiation *Photochem. Photobiol. Sci.* **14** 19–52

[15] Zerefos C, Contopoulos G and Skalkeas G (ed) 2009 Twenty years of ozone decline *Proc. of the Symp. for the 20th Anniversary of the Montreal Protocol* (Dordrecht: Springer)

[16] Gies P, Roy C, Javorniczky J, Henderson S, Lemus-Deschamps L, Driscoll C 2004 Global solar UV index: Australian measurements, forecasts and comparison with the UK *Photochem. Photobiol.* **79** 32–9

[17] Utrillas M P, Marín M J, Esteve A R, Estellés V, Gandía S, Núnez J A, Martínez-Lozano J A 2013 Ten years of measured UV Index from the Spanish UVB radiometric network *J. Photochem. Photobiol. B* **125** 1–7

[18] Krzyścin J W, Sobolewski P S, Jarosławski J, Podgórski J, Rajewska-Więch B 2011 Erythemal UV observations at Belsk, Poland, in the period 1976–2008: data homogenization, climatology, and trends *Acta Geophys.* **59** 155–82

IOP Publishing

Sun Protection
A risk management approach
Brian Diffey

Chapter 4

Deleterious effects of solar UV radiation exposure on the skin

The normal responses to UV radiation can be classed under two headings: acute effects and chronic effects. An acute effect is one of rapid onset and generally of short duration, as opposed to a chronic effect which is often of gradual onset and long duration. These effects should be distinguished from acute and chronic exposure conditions which refer to the length of the UV exposure. The skin is the critical organ resulting from solar UV radiation exposure, although eye damage can also result.

The skin covers the entire external surface of the human body and is the principal interface between our body tissues and the external environment. It serves as a protective barrier that prevents internal tissues from exposure to temperature extremes, toxins, bacteria, trauma, and UV radiation. Other functions of the skin include thermoregulation, sensory perception, and immunologic surveillance. The structure of human skin is shown in figure 4.1.

The skin comprises an outer layer of epidermis, which consists of layers of epithelial cells that are continually replaced by proliferation from the deeper epidermal layers by division of cells in the basal layer. Cells from the basal layer migrate towards the surface and in doing so they become flatter and the nuclei become smaller, until the most superficial epidermal layer—the stratum corneum—is composed of flat cells without nuclei.

The basal cells are separated from the deeper tissues by the dermo-epidermal junction. Below this is the dermis, connective tissue carrying blood vessels and nerves. The skin also contains hair follicles, sebaceous glands and sweat glands.

The thickness of the epidermis over much of the body surface is in the range 30–100 μm although on areas such as palms and soles, the epidermal thickness is around 400 μm. Full thickness skin is typically in the range 1–2 mm.

doi:10.1088/978-0-7503-1377-3ch4

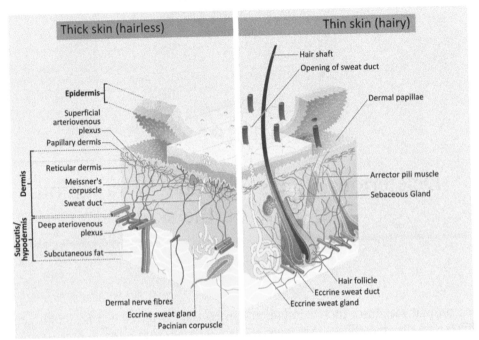

Figure 4.1. A schematic cross section of human skin. This image has been obtained by the author from the Wikimedia website where it was made available by Madhero88 and M Komorniczak under a CC BY-SA 3.0 licence. It is included within this article on that basis. It is attributed to Madhero88 and M Komorniczak.

4.1 Pathways to harm from solar UV exposure

From entering the skin to causing biological and clinical effects, UV radiation has to initiate a number of processes illustrated schematically in figure 4.2. These pathways encompass a variety of seemingly unconnected areas of knowledge ranging from climatology, optical physics, photochemistry, cellular and molecular biology through to clinical medicine and pathology.

The climatology of solar UV has been covered in chapters 2 and 3. Here we will review the remaining pathways that lead to harm.

4.1.1 UV penetration into skin

Optical radiation incident on the skin may be reflected at the skin surface due to a change in refractive index between air and stratum corneum, absorbed by chromophores in the epidermis or dermis, scattered by cell organelles in the epidermis or collagen in the dermis, or transmitted to deeper tissues.

The reflection of light from the surface of the skin is always between 4% and 7% for both black and white skin as a result of the difference in refractive indices between stratum corneum and air. UV radiation that enters the skin will be scattered mainly in a forward direction in the epidermis due to Mie scattering by cell organelles, e.g. melanosomes, which have dimensions of the order of the wavelength

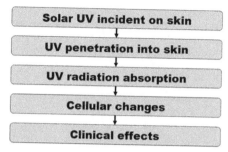

Figure 4.2. Pathways to harm from solar UV radiation.

of optical radiation. In the dermis, however, scattering is much more isotropic and radiation re-emitted from skin *in vivo* comes largely from the dermis. Dermal scattering increases rapidly with decreasing wavelength in an approximate manner predicted by Rayleigh scattering in which the probability of scattering varies inversely with the fourth power of the wavelength. Consequently, dermal scattering determines principally the depth to which different wavelengths of optical radiation penetrate the dermis.

It is evident from figure 4.1 that the skin is a complex structure that includes derivatives such as hair follicles, sweat glands and sebaceous glands. The skin contains irregularly shaped structures whose sizes are of the order of the wavelength of optical radiation. Consequently, a rigorous theoretical treatment of skin optics is formidable, although attempts to model the propagation of UV radiation in the skin have been made, largely based on radiative transfer theory.

There have been studies on the transmission of UV radiation through excised epidermis, but hardly any studies have reported on the distribution of radiation within the skin. The average transmission of UV radiation measured through the excised epidermis taken from the lower back of Caucasian and Asian participants that have not been exposed to sunlight or artificial sources of UV for several months is illustrated in figure 4.3.

Note the lower transmission through Asian skin compared with white skin, which is the result of a higher concentration of the pigment melanin.

The application of topical agents to the skin surface can alter its optical properties and so selectively increase or decrease radiation penetration to critical targets in the epidermis.

4.1.2 UV radiation absorption

From entering the skin to causing biological and clinical effects, optical radiation has to initiate photochemical processes.

When a molecule absorbs a photon of UV radiation, its electronic structure will change such that it reacts differently with other molecules. Each type of molecule in the skin, for example an amino acid or a nucleotide, absorbs a unique combination of wavelengths and is referred to as a *chromophore*. These differences in absorption characteristics between biomolecules underlie the diverse effects produced when skin is exposed to different wavelengths of optical radiation.

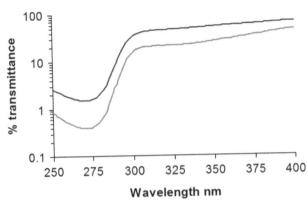

Figure 4.3. Average transmission of UV radiation measured through the excised epidermis taken from the lower back of 16 Caucasian subjects (blue curve) and 4 Asian subjects (red curve).

The energy that is absorbed by the chromophore can result in photochemical changes in the absorbing molecule itself, or in an adjacent molecule. Alternatively, the absorbed energy may be dissipated as heat, or as radiation of lower energy producing fluorescence or phosphorescence, in order to return the molecule to its ground state.

The basic laws of photochemistry
 - The First Law of Photochemistry (*Grotthus–Draper law*) states that radiation must be absorbed for photochemistry to occur. If a UV photon of a particular wavelength is not absorbed by a system, no photochemistry will occur, and no photobiological effects will be observed.
 - The Second Law of Photochemistry (*Stark–Einstein law*) states that for each photon of radiation absorbed by a system, only one molecule is activated in the primary step of a photochemical process.
 - The Bunsen–Roscoe Law of Reciprocity states that a photochemical effect depends on dose (intensity × time) and not dose rate (intensity). Reciprocity holds when the observable response depends only on the total administered radiant exposure (commonly referred to as dose) and is independent of the two factors that determine total dose, that is, irradiance and exposure time. Whilst this law applies to 'test tube' photochemistry, in skin photobiology patterns of exposure may be important, especially for melanoma (see section 4.4.6), that implies a failure of the reciprocity law.

Absorption spectrum
The structure of a molecule strongly influences which wavelengths of UV and/or visible radiation are absorbed. The chemical structure also determines the probability of absorption of photons at each wavelength, i.e. how much radiation is absorbed as a function of wavelength. A plot of the probability of absorption of photons against the wavelength is the *absorption spectrum*.
Examples of absorption spectra for some of the chromophores in skin are shown in figure 4.4.

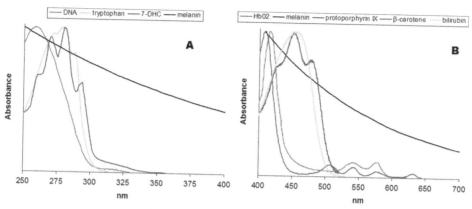

Figure 4.4. Absorption spectra of chromophores in skin.

The aromatic amino acids in proteins, particularly tryptophan and tyrosine, absorb UVB radiation. Tryptophan is responsible for most of the absorption of UV radiation by proteins. The purine and pyrimidine bases in DNA and RNA are also important UVB-absorbing chromophores for cutaneous responses. 7-dehydrocholesterol (7-DHC), which is involved in vitamin D synthesis (see section 2.3.1) also absorbs in this waveband (figure 4.4A).

Endogenous chromophores that absorb in the visible region include β-carotene, oxyhaemoglobin (HbO_2), and bilirubin. Melanin is unique since it absorbs throughout the UVB, UVA and visible wavebands and, unlike other biomolecules, shows no absorption maximum (figure 4.4B).

The spectra in figure 4.4 are intended to show only the distribution of wavelengths absorbed by various chromophores and not their relative absorption of UV and visible light in skin. The amount of radiation actually absorbed by these molecules in skin is related to the amount of each chromophore present. For example, DNA absorbs much more of the UVB radiation incident on skin than 7-DHC because it is present at much greater concentration.

Action spectrum

The wavelength dependence of a given photobiological effect is expressed by its *action spectrum*, which depends principally on the absorption spectrum of the chromophore and the optical properties of skin. Conventionally in skin photobiology, the reciprocal of the dose required to produce a given end-point is plotted against wavelength to obtain an action spectrum. Action spectroscopy studies show that UVB is much more effective than UVA for most endpoints studied in human skin including erythema and skin cancer.

4.1.3 Cellular changes

The main cell populations in the epidermis are keratinocytes, melanocytes and Langerhans cells, with fibroblasts and mast cells in the dermis. Some of the changes resulting in these cells following absorption of UV radiation, primarily in the UVB

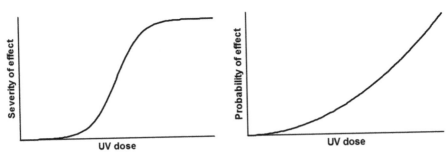

Figure 4.5. Deterministic (left) and stochastic (right) effects of UV exposure.

range, include the formation of cyclobutane pyrimidine dimers in DNA, of which more than 75% are thymine dimers with the remainder being pyrimidine [6–4] pyrimidone photoproducts, lipid effects and protein effects. Cellular adaptive responses to UV damage include melanocyte tanning response, cell cycle arrest, repair and, when damage is beyond repair, apoptosis (see section 8.4).

4.1.4 Clinical changes

The clinical effects of UV exposure can be either *deterministic*, where the magnitude of the effect is related to exposure and a threshold dose is possible, or *stochastic* in which the probability of the effect is related to exposure and there is no threshold dose (figure 4.5).

Erythema is an example of a deterministic effect and squamous cell skin cancer is a stochastic process. These, and other, effects are discussed in detail in the remainder of the chapter.

4.2 Erythema

Erythema is the commonest and most obvious effect of UV exposure. When the source of the UV exposure is the Sun, the reaction is referred to as sunburn.

Erythema is an acute injury following excessive exposure to UV radiation. The redness of the skin that results is due to an increased blood content of the skin by dilatation of the superficial blood vessels in the dermis. Half an hour of midday summer sunshine in temperate latitudes on unacclimatised white skin is normally sufficient to result in a subsequent mild reddening of the skin. Following this degree of exposure erythema may not appear for about 4–6 h, although exposing the skin for increasing periods to strong summer sunshine progressively shortens the time before the appearance of erythema, lengthens its persistence, and increases its intensity.

Erythema is associated with the classic signs of inflammation, which are redness and warmth, with high doses of UV resulting in oedema, pain, blistering, and, after a few days, peeling. Development of new pigmentation is much slower and takes several days following exposure.

The mechanism of erythema production following UV exposure is poorly understood. It is known that erythema is mediated, at least in part, by the release from the epidermis of pharmacologically active compounds, such as prostaglandins, which diffuse to act on dermal blood vessels. The erythemal response may also be related to the DNA damaging effects of UV radiation.

4.2.1 Factors influencing the development of erythema

Skin colour is an important factor in determining the ease with which the skin will experience erythema following UV exposure. Other phenotype characteristics that may influence the susceptibility to erythema are hair colour, eye colour and freckles.

In 1975, Dr Thomas Fitzpatrick proposed a method of predicting human UV sensitivity on the basis of a personal history of response to 45–60 min of exposure to midday sun at a latitude of 40 °N in early summer [1]. This method utilises information on both erythema and pigmentation collected in an interview to place individuals in one of the six phototypes, as shown in table 4.1.

People with the lowest phototypes (and the lowest melanin content in the skin) are the most UV sensitive and those with the highest phototypes (and the highest melanin content are the least UV sensitive (figure 4.6).

4.2.2 Minimal erythema dose (MED)

Individual sensitivity to UV-induced erythema varies widely in skin of different complexion and is associated with the cutaneous melanin content, phototype and race/ethnicity. This predisposition is usually inferred from the *minimal erythema dose* (MED). This value is determined by exposing adjacent areas of skin to increasing doses of UV radiation (usually employing a geometrical series of dose increments)

Table 4.1. Six phototypes of human skin.

Phototype	Skin reactions to UV exposure	Examples
I	Always burns easily and severely (painful burn); tans little or not at all, and always peels	People most often with fair skin, numerous freckles, red hair, blue eyes; unexposed skin is pinkish-white
II	Usually burns easily and severely (painful burn); tans minimally, also peels	People most often with fair skin, some freckles, blond or reddish-brown hair, blue, hazel or brown eyes; unexposed skin is white
III	Burns moderately and tans about average	Normal average Caucasian; few freckles may be present, unexposed skin is white
IV	Burns minimally, tans easily	People with naturally white or light brown unexposed skin, dark brown hair, dark eyes.
V	Rarely burns, tans easily and substantially	People with naturally brown skin, dark hair
VI	Never burns and tans profusely	People with naturally dark brown or black skin, dark hair

Figure 4.6. Sun-reactive skin types (image courtesy of American Laser Med Spa, Amarillo TX 79106).

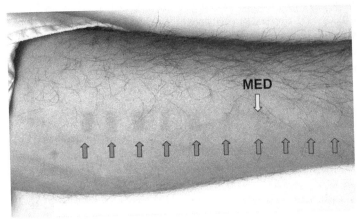

Figure 4.7. A series of increasing exposures to UV radiation on the forearm. The MED is indicated by the yellow arrow.

and recording the lowest dose of radiation to produce erythema at a specified time, usually 24 h, after irradiation, as illustrated in figure 4.7.

The visual detection of erythema is subjective and is affected by unrelated factors such as viewing geometry, intensity and spectral composition of ambient illumination, colour of unexposed surrounding skin, as well as the experience and visual acuity of the observer. The difficulty in judging a minimal erythema response accurately is reflected by the varying definitions proposed for this value: these range from the dose required to initiate a just perceptible erythema, to that dose which will just produce a uniform redness with sharp borders. The former end-point has been shown to be more reproducible and less prone to inter-observer differences than the latter.

The MEDs determined on previously unexposed buttock skin using solar simulated radiation in a total of 345 individuals of sun-reactive skin types I (78 subjects), II (230 subjects), III (27 subjects), and IV (10 subjects) [2] are shown in figure 4.8 as a box-and-whisker plot. The horizontal lines either side of the rectangle represent the lower and upper quartiles and the horizontal red line inside the rectangle indicates the median value, with the minimum and maximum values indicated by the error bars.

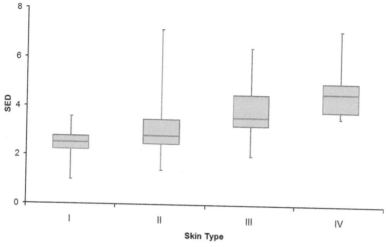

Figure 4.8. A box-and-whisker plot showing the MEDs measured in different skin types (data supplied by Professor Antony Young from reference [2]).

We can see that whilst the median MED increases with skin type, as to be expected, the range of data within each skin type indicates that there is a large overlap between the different skin types, meaning that subjective skin type assessment is a poor predictor of erythema sensitivity. The median MEDs for skin types I, II, III and IV are 2.5, 2.8, 3.6 and 4.6 SED, respectively.

An alternative method to visual determination of the MED is to use a reflectance colorimeter, which expresses skin colour in the CIE $L^*a^*b^*$ colour space coordinates. The three coordinates represent the luminance, or brightness, of the colour ($L^* = 0$ yields black and $L^* = 100$ indicates diffuse white), its position between red and green (a^*, negative values indicate green and positive values indicate red) and its position between yellow and blue (b^*, negative values indicate blue and positive values indicate yellow).

From these values, skin colour can be expressed by the individual typology angle, ITA°, determined as:

$$ITA = \{\arctan[(L^* - 50)/b^*]\} \times 180/\pi$$

Individuals are classified according to their ITA, as follows:

ITA° > 55°	Very light
55° ⩾ ITA° > 41°	Light
41° ⩾ ITA° > 28°	Intermediate
28° ⩾ ITA° > 10°	Tan
10° ⩾ ITA° > −30°	Brown
−30° ⩾ ITA°	Dark

In the same reference cohort of subjects, the MED would be determined and a relationship constructed between MED and ITA°. The MED in subsequent subjects would be obtained by measuring their skin colour, determining the ITA°, and estimating the equivalent MED from the relationship established in the reference cohort.

However, as a rough and ready alternative we can equate skin complexions very light, light, intermediate and tan with skin types I, II. III and IV, respectively, and crudely estimate the MED (expressed in SED) in unexposed white skin as:

$$MED = 5.5 - 0.05 \times ITA°$$

4.2.3 Anatomical variation in erythemal sensitivity

Erythemal response has been measured on different areas of the skin such as face, neck and trunk, upper and lower back, buttocks, abdomen, and extremities. The face, neck and trunk are two to four times more sensitive than the limbs.

The differences in UV sensitivity of different body sites reflect site-specific variations in skin colour i.e. melanin content, and this is found in both Asian and Caucasian skin. The area of the back above the waistband is most commonly used for testing sunscreens.

4.2.4 Effect of gender and age on erythemal sensitivity

There is no difference in sunburn susceptibility between sexes and quantitative studies of erythemal sensitivity do not indicate changing sensitivity with age.

4.2.5 Time course of sunburn

Depending on the degree of solar UV exposure, skin redness is first visible within 1–8 h after irradiation and maximum intensity is seen between 8–24 h, after which there is a gradual decline in intensity.

Although there is an apparent latent period of a few hours between going out into the Sun and appearance of erythema, sensitive instrumental studies have confirmed that vasodilatation is occurring as soon as irradiation commences. The reason we do not perceive immediate skin reddening is due to the insensitivity of the eye and not to any real latent period.

4.2.6 Spectral response

The ability of UV radiation to produce erythema in human skin is highly dependent upon the radiation wavelength, and is expressed by the erythema action spectrum. This action spectrum is normally obtained by using a broad-band optical radiation source, such as a xenon arc lamp, in conjunction with a monochromator incorporating a diffraction grating as the dispersion device. Small areas of skin, usually on the back, are irradiated with increasing doses at a range of wavelengths and at each wavelength, the MED is recorded. The reciprocal of the MED, normalised to unity at the peak value, is plotted against wavelength to obtain the action spectrum.

Erythema action spectra have been the subject of experimental and theoretical interest for almost 100 years. The International Commission on Illumination (CIE) first considered the adoption of a standard erythemal curve in 1935. However, this curve differed appreciably from erythema action spectra determined in the second half of the 20th century, particularly at wavelengths less than 300 nm. As a consequence, a reference action spectrum, based upon a statistical analysis of the results of published studies in this period, was proposed in 1987 [3], and officially adopted by the CIE in 1998 [4]. It is represented by relatively simple functions over three clearly defined spectral regions:

$$S(\lambda) = 1.0 \qquad\qquad 250 \leqslant \lambda \leqslant 298 \text{ nm}$$
$$S(\lambda) = 10^{0.094(298-\lambda)} \qquad 298 < \lambda \leqslant 328 \text{ nm}$$
$$S(\lambda) = 10^{0.015(140-\lambda)} \qquad 328 < \lambda \leqslant 400 \text{ nm}$$

This reference action spectrum is used whenever there is a need to calculate an erythemal-effective dose, usually expressed by the standard erythema dose (section 3.2.1).

The original 1935 CIE erythema action spectrum is compared in figure 4.9 with the current CIE 1998 reference spectrum and that obtained in 1982 from an experimental study using a 5000 W Xe–Hg compact arc lamp optically coupled to a holographic grating monochromator in which the MED was determined at several wavelengths between 250 and 435 nm at 24 h after exposure [5]. Erythema was not observed at the highest exposure dose of 3×10^6 J m^{-2} at 435 nm and so the relative sensitivity lies below the data point, as indicated by the arrow in figure 4.9. This does not mean that skin will not develop delayed erythema in response to visible light, simply that the exposure dose given was insufficient to induce a reaction.

Note that a logarithmic axis is used to plot relative sensitivity reflecting the strong dependence on UV wavelength.

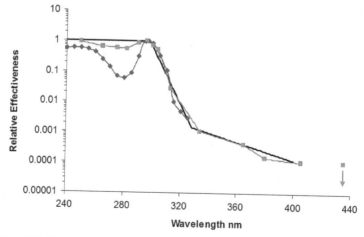

Figure 4.9. The CIE (1998) reference action spectrum for erythema in human skin [4] (black), an experimentally-determined erythema action spectrum [5] (red), and the CIE (1935) action spectrum (blue).

That the erythema action spectra in figure 4.9 cease at 400 nm begs the question whether delayed erythema can be induced by longer wavelengths. There are very few experimental studies on cutaneous responses to visible light and although broad band visible light has been shown to produce immediate erythema that fades within two hours or so and thought to be due to a thermal, and not photochemical, response [6], there are no data on the induction of delayed (24 h) erythema in normal skin following visible light irradiation, primarily due to insufficient radiant exposure in the studies conducted to date.

However, as we shall see in section 4.3.1, in the photosensitivity disorder chronic actinic dermatitis (CAD), erythemal responses at 24 h following exposure with an irradiation monochromator have been observed at wavelengths in the visible region up to 600 nm.

Furthermore, the action spectrum in the UV region for induction of an erythemal response at 24 h in CAD has been shown to be the same shape as that for normal sunburn in individuals with fair skin [7]. This suggests that an endogenous chromophore(s), responsible for initiation of the abnormal reaction to light in CAD may be the same as, or similar to, that/those responsible for erythema in normal skin. It could well be, therefore, that, at sufficient exposure doses, radiation at different wavelengths throughout the visible region is capable of inducing delayed erythema in normal skin.

4.2.7 Dose response

The MED is a single point on the dose–response curve and an apparent threshold measurement. It provides no information about the effect of higher doses, and consequently may fail to reveal significant differences between responses. Because of this, a number of methods have been described to quantify erythema.

An attempt to quantity the degree of redness at each site by eye by assigning grades such as +−, +, ++, +++ has several drawbacks. It is subjective and, although under ideal viewing conditions it is possible to detect small differences in erythemal intensity between adjacent irradiated sites, the eye is poor at estimating the exact value of the difference. Also, linearity is often assumed; that is to say, the erythemal increments between grades are equal, e.g. between + and ++, and between ++ and +++, whereas many ordinal scales have a skewed distribution.

The earliest attempt to quantify reproducibly the degree of redness of UV-induced erythema was by use of colour comparison charts. Again, although subjective, this method has an advantage over the visual grading of erythema as the eye performs well in a null system such as the matching of two colours. The main drawback of this system is that the colour standards do not have the general appearance of skin in that they lack the details of hairs, surface texture, translucency and irregularities of colour, which are conspicuous on visual examination.

A similar approach is to use red-coloured optical filters in which a series of photographic filters with a high transmittance for red light and decreasing transmittance for blue–green light is used and to note which of the filters just causes the erythema to disappear.

Instrumental techniques for quantifying erythema include reflectance spectropho- tometry, colorimetry, and computer-assisted digital image analysis. It may be helpful to understand the principle of reflectance spectrophotometry by reference to figure 4.10.

The left hand image in figure 4.10 shows a series of UV-induced erythematous sites resulting from increasing doses of UV going from bottom to top. Under ambient illumination the sites appear red, the reason being that the blue and green components of the incident white light are preferentially absorbed by oxyhaemo- globin in the increased blood content of the sites (see figure 4.4B), whereas the red light component is unaffected. As a consequence, the spectrum of light reflected to the observer's eye is deficient in photons in the blue and green wavebands relative to the red waveband.

This is clearly seen in the central image in figure 4.10 where the subject has been photographed under green light illumination resulting in increasingly darker spots with increasing erythema. However, under red light illumination (right hand image) the erythematous sites seem to disappear. This means that there is no more red light reflected from sites that show strong erythema than unirradiated white skin. So it might be more correct to say to someone who is sunburnt that their skin has not gone red but less green.

By measuring the reduction in intensity of green light reflected from different erythematous sites, relative to reflection of red light from the corresponding site, we can derive a measure of 'redness' normally termed the erythema index. Several different instrumental methods have been used since the 1920s to evaluate erythema

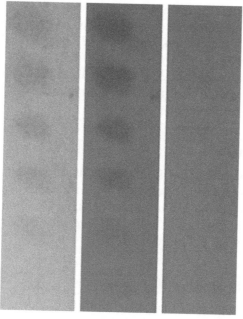

Figure 4.10. A series of increasing exposures to UV radiation illuminated by white light (left), green light (centre) and red light (right).

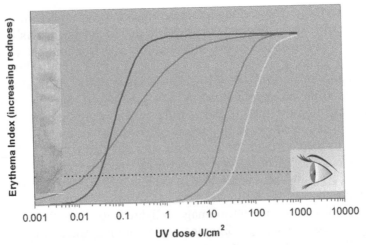

Figure 4.11. Typical UV erythema dose–response curves for UVC (blue curve), UVB (green curve), UVA (yellow curve) and sunlight (red curve). The broken horizontal line indicates the approximate limit of visual sensitivity and corresponds to the MED.

by exploiting the strong absorbance of blue and green light and the weak absorbance of red light and there are a number of commercial instruments that measure an erythema index (and sometimes a complimentary melanin index).

After exposing the skin to increasing doses of UV radiation, as illustrated in figure 4.10, followed by measuring the erythema index at each irradiated site, we can construct a dose–response curve (figure 4.11) that follows an S-shaped curve, where the dose is on a log scale, as often seen with pharmacological data.

The curve consists of three regions: (a) little or no response is seen at low doses; (b) the middle part of the curve is approximately linear; (c) the rate of increase of erythema falls at high doses.

From figure 4.11 we see very clearly that the slope of the mid-part dose–response curve is much shallower for UVC radiation than for either UVB, UVA or erythema induced by sunlight, i.e. sunburn.

4.3 Photosensitivity

Photosensitive diseases are not a 'normal' response to sunlight, rather they represent a pathological response of the skin to specific wavebands of the solar spectrum.

The diseases, termed the *photodermatoses*, may be immunologically-mediated, induced as a result of chemicals or drugs, DNA repair-deficient photodermatoses, or skin diseases aggravated, but not caused, by sunlight.

4.3.1 Immunologically-mediated photodermatoses

Polymorphic light eruption (PLE), actinic prurigo, hydroa vacciniforme, chronic actinic dermatitis and solar urticaria belong to this group of idiopathic photo-dermatoses, of which the most common is PLE.

Polymorphic light eruption

As the name suggests there may be a variety of morphological lesions on light-exposed areas, for example, erythema, eczema, papules, plaques or blisters. The latent period between exposure and appearance of the rash lies between a few hours to two days. Attacks generally start in the spring and continue through until autumn. There is, however, a tendency in some patients for the eruptions to subside in the summer. The site of eruption depends upon sunlight exposure of unclothed areas, although the most favoured sites are the backs of the hands, the front of the neck and the face.

The age of onset can be very variable although PLE usually appears in the second and third decades, and it is more common in women than in men. Once present, the disease seems to be recurrent and can last an indefinite number of years. Perhaps surprisingly, the prevalence of PLE is inversely related to latitude. Patients with PLE generally have a normal sunburn response.

Chronic actinic dermatitis

Chronic actinic dermatitis is a persistent and distressing condition occurring mainly in older people, especially men, who commonly have a long history of chronic skin disease such as eczema. The rash is not necessarily confined to sun-exposed sites and lesions can extend onto covered areas. There is a markedly increased abnormal erythemal reactivity over a wide spectrum extending from the UVB region, through the UVA and not uncommonly to about 600 nm.

Solar urticaria

Unlike the other idiopathic photodermatoses, solar urticaria evolves during or within a few minutes of exposure and usually resolves within half an hour. Erythema localised to the exposed skin site appears first and may be accompanied by a sensation of itching or burning. This is followed by the development of a weal which reaches its greatest extent in about 5–10 min There does not appear to be a definitive action spectrum for the disease; different patients have reacted in the UVB, UVA, and visible spectrum.

4.3.2 Drug and chemical photosensitivity

This group of conditions may be due to endogenous or exogenous chemicals.

Porphyria

Porphyria is the collective name given to a rare group of hereditary disorders associated with specific enzyme deficiencies in haem biosynthesis and so is an endogenous chemically-induced photosensitivity. A feature of the disease is the excessive production of porphyrins, which appear in the urine either as red pigments or as colourless compounds that darken on exposure to light. One of the clinical symptoms of porphyria is skin photosensitivity, which may be evident at an early age.

Porphyria photosensitivity may manifest itself as subjective symptoms, often of a burning sensation, during or soon after exposure to sunlight, and skin changes such as erythema or oedema which can appear within an hour or up to 18 h after exposure.

Exogenous drug and chemical photosensitivity
Exogenous chemicals that cause photosensitisation may enter the body by ingestion, injection or absorption through the skin.

Topical substances such as tar or synthetic dyes give symptoms such as erythema or oedema which can occur within minutes of exposure to the Sun. Reactions to topical sensitisers are classed as either phototoxic or photoallergic. Some sunscreens can produce a photoallergic reaction although the incidence is probably very low.

Orally administered photosensitisers, such as therapeutic drugs, result in a phototoxic reaction that, theoretically, can arise in any subject given sufficient exposure to light and chemicals. Photosensitising drugs include antibiotics, anti-fungal agents, diuretics, non-steroidal ant-inflammatory drugs and psychoactive drugs.

4.3.3 DNA repair-deficient photodermatoses

These diseases, which include xeroderma pigmentosum, Cockayne Syndrome and trichothiodystrophy are extremely rare autosomal recessive genetic conditions that exhibit a number of different clinical features, one of which is abnormal sensitivity to sunlight.

4.3.4 Photoaggravated skin diseases

These represent a very heterogeneous group of conditions that share only one common feature—in some patients they can be exacerbated by exposure to sunlight or to artificial sources of UV radiation. Skin diseases that fall into this group include acne, psoriasis, lupus erythematosus, and herpes simplex and other viral rashes.

4.4 Skin cancer

The three common forms of skin cancer, listed in order of increasing seriousness are basal cell carcinoma (BCC), squamous cell carcinoma (SCC) and malignant melanoma (MM). Around 90% of skin cancer cases are of the non-melanoma variety (BCC and SCC) with BCCs being approximately 3–4 times as common as SCCs.

There are other less common types of skin cancer that make up only about 1% skin cancers. These are Merkel cell carcinoma, Kaposi's sarcoma and T cell lymphoma of the skin (sometimes called primary cutaneous lymphoma).

4.4.1 Clinical features

Basal cell carcinoma, which makes up around 70% of all skin cancers and is sometimes referred to as a rodent ulcer, can occur in a number of different subtypes that include nodular, superficial, morphoeic and pigmented; about a half of BCCs are the nodular type. It is very rare for basal cell skin cancers to metastasise. An example of a BCC is shown in figure 4.12.

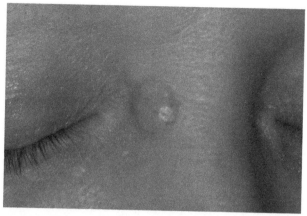

Figure 4.12. A basal cell carcinoma on the right side of the bridge of the nose (courtesy of Dr James Langtry).

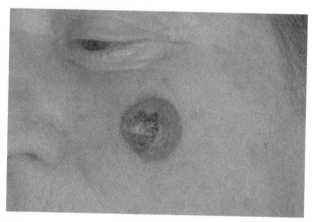

Figure 4.13. A squamous cell carcinoma on the left cheek (courtesy of Dr James Langtry).

Squamous cell skin cancer is generally faster growing than BCCs and constitutes about 20% of skin cancers. Like BCCs, SCCs develop in areas that are habitually exposed to the Sun such as the face (figure 4.13), neck, and on the back of hands and forearms. SCCs seldom spread and if they do, it is most often to the deeper layers of the skin. Rarely they can spread to nearby lymph nodes and other organs causing metastatic disease.

Malignant melanoma is the most serious skin cancer as it is often lethal, although 90% of people who develop melanoma can expect to survive for 10 years or more. Most melanomas appear as a pigmented lesion with an irregular border and are frequently multicoloured mixtures of black, brown and pink (figure 4.14).

There are four main types of malignant melanoma: superficial spreading melanoma (the most common type in Caucasians), nodular melanoma (the most rapidly growing and aggressive type), acral melanoma (occurs on palms and soles) and lentigo maligna melanoma (slow growing and confined to the skin of the elderly).

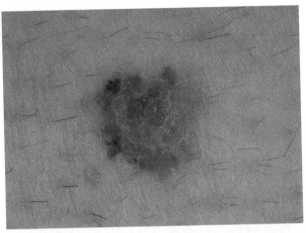

Figure 4.14. A superficial spreading malignant melanoma of the left leg (courtesy of Dr James Langtry).

An important factor in determining the progression of malignant melanoma is the Breslow thickness, which measures (in millimetres) the distance between the upper layer of the epidermis and the deepest point of tumour penetration. The thinner the melanoma, the better the chance of a cure. The least serious melanoma is *in situ* (non-invasive), which remains confined to the epidermis. Thin tumours are less than 1 mm in Breslow thickness, intermediate tumours are 1 to 4 mm, and thick melanomas are greater than 4 mm. The superficial spreading malignant melanoma shown in figure 4.14 has a Breslow thickness of 0.5 mm, i.e. a thin melanoma.

In general, the number of cases of *in situ* melanoma in a given time period is roughly similar to the number of invasive melanomas.

4.4.2 The world burden of skin cancer

Skin cancer is by far the most common cancer diagnosed in most western countries. Cases of basal and squamous cell skin carcinomas are not recorded by cancer registries in many countries and so incidence data of these tumours in different countries are not available. However, cases of malignant melanoma are recorded and the incidence of this disease varies by more than ten-fold around the world (figure 4.15).

It is clear from this map that melanoma is much more prevalent in countries where the indigenous population is predominantly white. Regions of the world in the tropical belt roughly between 25 °N and 25 °S generally have a much lower incidence even though insolation is higher. The predominantly darker skins of people living in these regions is attributed to this observation.

4.4.3 Age dependence of skin cancer

Age-specific incidence rate
Table 4.2 gives the number of cases of malignant melanoma in 5 year age bands diagnosed in males in England in 2011, along with the population of males in the respective age bands. The fourth column calculates the age-specific rate expressed as

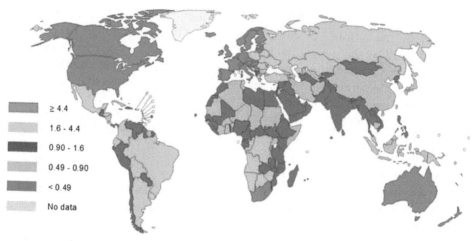

Figure 4.15. Estimated age standardised incidence rates of melanoma (both sexes) in 2012 [8]. The incidence rate bands refer to the number of new cases arising per 100 000 persons in 2012. Reproduced with permission from Ferlay J, Soerjomataram I, Ervik M, Dikshit R, Eser S, Mathers C, Rebelo M, Parkin DM, Forman D, Bray, F. GLOBOCAN 2012 v1.0, Cancer Incidence and Mortality Worldwide: IARC CancerBase No. 11 [Internet]. Lyon, France: International Agency for Research on Cancer; 2013. Available from: http://globocan. iarc.fr, accessed on 5 October 2017.

Table 4.2. Computation of age-specific incidence rates and age-standardised incidence rates (malignant melanoma, England, males, 2011).

Age band i	No. of cases n_i	Population at risk y_i	Age-specific incidence per 10^5 population $10^5 \times n_i/y_i$	Standard world population* w_i	Expected cases in standard population $10^3\, n_i w_i\, /\, y_i$
0–4	1	1 703 700	0.1	8.86	0.01
5–9	1	1 530 500	0.1	8.69	0.01
10–14	2	1 570 300	0.1	8.60	0.01
15–19	20	1 701 100	1.2	8.47	0.10
20–24	40	1 815 500	2.2	8.22	0.18
25–29	103	1 825 900	5.6	7.93	0.45
30–34	121	1 765 300	6.9	7.61	0.52
35–39	157	1 755 700	8.9	7.15	0.64
40–44	285	1 921 900	14.8	6.59	0.98
45–49	361	1 926 100	18.7	6.04	1.13
50–54	422	1 699 000	24.8	5.37	1.33
55–59	443	1 484 800	29.8	4.55	1.36
60–64	688	1 552 000	44.3	3.72	1.65
65–69	634	1 242 600	51.0	2.96	1.51
70–74	675	963 600	70.0	2.21	1.55
75–79	614	761 100	80.7	1.52	1.23
80–84	475	524 200	90.6	0.91	0.82
85+	398	389 900	102.1	0.63	0.64
Total	**5440**	**26 133 200**	**20.8**	**100**	**14.1**

*WHO World Standard [9].

the number of cases per 100 000 of the population at risk, which is simply 100 000 multiplied by the number of cases diagnosed in a particular age band (second column) divided by the population at risk (third column).

Similarly, we calculate the age-specific rates for males and females in England in 2011 who were diagnosed with BCC and SCC. These are plotted in figure 4.16 and we see clearly that age is a major determinant of incidence.

The risk of epithelial cancers, which comprise 90% of all cancers worldwide, increases approximately as a power function of age [10]. This can be expressed mathematically as:

$$I(t) \propto t^{\alpha}$$

where $I(t)$ is the incidence at age t and α is a constant termed the *age exponent*.

We see from figure 4.16 that when these data are plotted on a log–log axis, the curves are approximately linear with slopes equal to the age exponent α for a given tumour and gender. Figure 4.16 shows that skin cancers are very rare in people under 20 years and that SCC increases more rapidly with age than BCC. The slopes calculated between ages of 25 to 84 for BCC are 4.8 and 4.0 in males and females, respectively. Corresponding slopes for SCC are 7.0 and 6.3.

The age dependence of melanoma is appreciably different, as shown in figure 4.17.

In contrast to most cancer types, malignant melanoma occurs relatively frequently at younger ages. For example, in England in 2011 around half of cases were diagnosed in people aged below 60.

The cumulative risk of a given cancer up to age 80 years (average life expectancy in England in 2011) is estimated according to:

$$\% \text{ cumulative risk} = 100 \times \left[1 - \exp\left(-5 \times \sum_{i=1}^{A} I_i / 100\ 000 \right) \right]$$

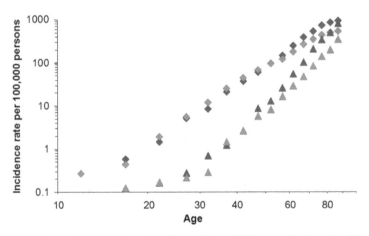

Figure 4.16. Age-specific incidence rates for BCC (diamonds) and SCC (triangles) in males (blue) and female (red) for England in 2011.

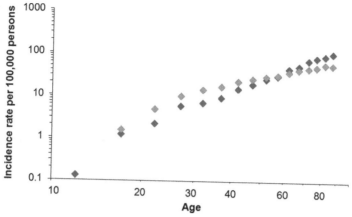

Figure 4.17. Age-specific incidence rates for malignant melanoma in males (blue) and female (red) for England in 2011.

where the summation is up to age-band A (i.e. 75–79 years), and I_i is the age-specific incidence rate in the ith age-band, each age band having a width of 5 years. This expression gives an approximate measure of the lifetime risk assuming no other causes of death are in operation. For skin cancers diagnosed in England in 2011, these estimates are:

	BCC		SCC		MM	
	Males	**Females**	**Males**	**Females**	**Males**	**Females**
Probability %	10.6	7.6	3.7	1.6	1.8	1.6
Chance of 1 in	9	13	27	61	56	61

Age standardised incidence rate
The age-standardised incidence rate is the summary rate that would have been observed, given the schedule of age-specific rates, in a population with the age composition of some reference population.

The calculation is illustrated in table 4.2 for melanoma incidence among males in England in 2011, and uses a standard based on the mean world population age structure projected for the period 2000–25 proposed by the World Health Organization [9]. Other standards are also used in comparing cancer incidence between countries, for example, based on a European standard population.

The crude rate per 100 000 per year is calculated as:

$$10^5 \sum_i n_i \bigg/ \sum_i y_i = 10^5 \times 5440/26\ 133\ 200 = 20.8$$

The age-standardised rate is given by:

$$10^3 \sum_i n_i w_i / y_i = 14.1$$

n_i are the number of cases reported among males in the ith age band during 2011, y_i is the population at risk in this age band, and w_i is the weighting factor for the standard world population. In this example, the age-standardised rate is two-thirds the crude rate. This is because the standard world population has proportionally fewer individuals in the older age groups than the corresponding English population, and the risk of disease (age-specific rates) is highest in the oldest age groups.

The number of recorded cases and the corresponding age-standardised rates for BCC, SCC and MM in England in 2011 are summarised below.

	BCC		SCC		MM	
	Males	Females	Males	Females	Males	Females
No of cases	33 485	28 174	13 312	8082	5440	5681
Age-standardised rate	77.1	57.0	27.1	12.3	14.1	14.7

We see that in 2011 there were 94 174 cases of skin cancer recorded out of a population of just over 53 million people.

Whilst skin cancer data for England for the year 2011 has been used to illustrate age-specific and age-standardised incidence rates, data from cancer registries around the world are available to estimate measures of incidence in different countries and time periods.

4.4.4 Trends in skin cancer incidence

The incidence of skin cancer, and especially malignant melanoma, has increased steadily over the past 50 years in predominately fair-skinned populations. The trends in incidence probably reflect changing prevalence of risk factors such as increased leisure time in sunny destinations, changing fashion and sunbed use, coupled with increased surveillance, early detection and changes in diagnostic criteria.

As an example, figure 4.18 illustrates the age-standardised rates in Australia (a country of high insolation) and the UK (a country of low insolation) over a 30 year period.

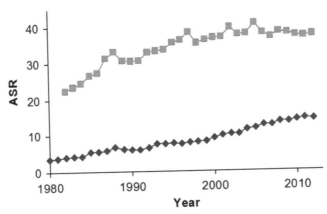

Figure 4.18. Trends in the age-standardised rate per 100 000 (ASR) of malignant melanoma in Australia (red squares) and Great Britain (blue diamonds).

Whilst the incidence is considerably higher in Australia compared with the UK, we see the incidence continues to rise in the UK, whereas the rate has stabilised since the mid-1990s in Australia. One reason might be the impact on incidence trends of skin cancer prevention activities that were first initiated in Australia from the 1960s but only towards the end of the 20th century in the UK.

Observed incidence data on melanoma over time are subject to the influence of many factors that include period effects and cohort effects.

Period effects result from external factors that equally affect all age groups at a particular calendar time. These could arise from a range of environmental, social or economic factors; for example, educational awareness and prevention campaigns or depletion of the ozone layer resulting in higher levels of ambient ultraviolet radiation. Also, methodological changes in outcome definitions, classifications, or method of data collection, such as increased surveillance, early detection and changes in diagnostic criteria, could also lead to period effects in data.

Cohort effects are variations resulting from the unique experience/exposure of a group of subjects (cohort) as they move across time leading to differences in the risk of a health outcome based on birth year. For example, following the widespread introduction of sunbeds for cosmetic tanning in the 1980s and their popularity amongst younger people, it would be expected that cohorts born after 1960 would be greater users of this form of UV exposure than cohorts born in earlier years.

4.4.5 Economic burden of skin cancer

In addition to causing illness and death, skin cancer is associated with an economic burden to society. In the United States, for example, skin cancer treatment is estimated to cost about $8.1 billion each year. In addition to direct medical costs, the annual cost associated with lost workdays and restricted-activity days are estimated to be around $100 million. An individual in the United States dying from melanoma loses an average of 20.4 years of potential life, compared with an average of 16.6 years for all malignant cancers and the annual productivity losses associated with these lost years is estimated to cost an additional $4.5 billion [11].

4.4.6 Association with sun exposure

Skin cancer is the most common human cancer and there is little dispute that chronic exposure to solar UV radiation is considered to be a major etiological factor for all three forms of this cancer. However, the evidence differs for BCC and SCC, collectively referred to as non-melanoma skin cancer (NMSC), and for melanoma. We consider first the evidence for NMSC.

Anatomical distribution
Numerous studies have shown that frequently-exposed body areas are the commonest sites for both BCC and SCC. For example, a large study from Queensland [12] found that most BCCs were on the head and/or neck (40%) and the trunk (34%), and that 33% of SCCs were on the head and/or neck with 35% on the upper limbs. To allow for differences in the skin surface area at different body sites, the relative tumour density

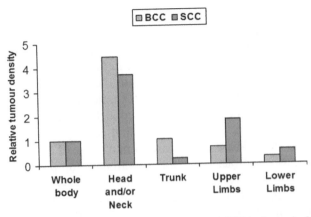

Figure 4.19. Relative tumour densities of BCC and SCC by body site [12].

for each body site was calculated by dividing the proportion of tumours at a site by the mean proportion of the skin surface area of that site (see figure 4.19).

When allowance is made for surface area, we see that for both BCC and SCC, the head and/or neck are much more prone to tumours than other body sites. Studies made comparing the geometry of sun exposure over the head and neck areas show a close association with sites of these tumours.

Unlike NMSC which predominates on sites of highest insolation (head and neck), melanoma occurs relatively more frequently on the trunk and legs. This distribution lends support to the *intermittent exposure* hypothesis discussed below.

Racial differences
People with white skins are much more likely to develop skin cancer than races with more marked pigmentation. This is illustrated in table 4.3 which shows the age-standardised rate of skin cancers in white people and black people living in the USA [13] and South Africa [14]. We see that melanoma and SCC are about 20 times more common in whites than blacks, and for BCC there is a 60-fold difference.

When skin cancer does occur in pigmented races it is not necessarily found on predominantly light exposed areas.

Phenotype
Genetic factors associated with a tendency to develop skin cancer are light eyes, fair complexion, light hair colour, tendency to sunburn and poor ability to tan. In other words, the risk of skin cancer decreases with increasing phototype (see figure 4.6).

There is an additional important risk factor for developing melanoma, which is the total number of naevi (moles) larger than 2 mm diameter.

Geographical distribution
People living in sunny countries such as Australia have a much greater risk of developing skin cancer than people with similar genetic backgrounds who live in much less sunny climates, such as the UK. For example, about 2 out of every

Table 4.3. Age-standardised rate of skin cancers in people with white skin and black skin.

Malignant Melanoma	Period	White skin	Black skin	Ratio
USA	1973–2013	16.2	0.7	22
South Africa	2000–09	18.1	1.1	16
BCC				
South Africa	2000–09	146.2	2.3	63
SCC				
South Africa	2000–09	50.7	2.5	20

3 people living in Australia can expect to develop at least one skin cancer during their lifetime, whereas the corresponding figure for the UK is about 1 in 10 people.

Whilst an inverse relationship between melanoma incidence and latitude of residence generally holds true, there are some inconsistencies. For example, in Europe the incidence is higher in Scandinavia than in Mediterranean countries. This inconsistency may reflect ethnic differences in constitutional factors as well as national customs and behaviours relating to sun exposure.

Occupation

People who work outdoors are more likely to develop non-melanoma skin cancers than indoor workers; for example, data collected in Sweden indicated that the risk in outdoor workers is about three times higher than for indoor workers.

Melanoma, on the other hand, is more common in professional and technical indoor workers than in those who work outdoors, such as farmers. Unlike all other cancers, melanoma is more likely in those of higher socioeconomic status. One reason that might account for this is that indoor workers, particularly those in higher socioeconomic groups, are likely to travel to sunny climates several times a year and so experience bursts of high sun exposure on skin that is covered up for much of the year. Because of the nature of their employment, outdoor workers are more likely to have skin that is acclimatised to sun exposure year-round.

Migration and critical period exposure

People born in Europe and who migrate to sunnier countries such as Australia or New Zealand after childhood have a risk of developing skin cancer of about one quarter of that of people of European descent born in those countries. However, arrival during childhood before the age of 10 years results in a comparable risk for BCC and melanoma.

This observation could suggest that skin cancer risk simply increases with length of residence in a sunny country, but can also imply that sun exposure early in life plays a crucial role in the lifetime risk of developing skin cancer.

Intermittent exposure hypothesis

Although the evidence from epidemiological studies indicates an association between melanoma and sunlight exposure, it does not appear that cumulative sun exposure explains the relationship, as it does for NMSC. Instead, there is evidence that intermittent sun exposure mainly from recreational activities, rather than cumulative or chronic exposure associated with occupation, is associated with increased risk of developing malignant melanoma. Several studies have established a history of sunburn as an important risk factor for melanoma development, although in these studies a potential for recall bias exists.

It may be that sunburn *per se* is not a direct risk factor for melanoma but could simply be a marker of a high sun exposure. People with albinism, an inherited disorder of melanin synthesis characterised by the reduction or complete absence of melanin pigment in the skin, hair and eyes, show an increased frequency of sunburn and NMSC compared with non-albino individuals but do not appear to be at an elevated risk of melanoma.

4.4.7 Non-solar risk factors for skin cancer

Immune suppression
Patients who have undergone organ transplantation, such as heart, lung and kidney, are given drugs to suppress their immune system so that it will not attack the donated organ as a foreign invader. One consequence of this is that immune suppression leads to a greater occurrence of skin cancers, especially SCC, resulting in more than 100-fold increased risk in transplant patients compared with the non-transplanted population.

HIV infection
People infected with the HIV virus have something like a two-fold increase in incidence rate of NMSC compared with a non-infected control group.

4.4.8 Action spectrum

Clearly, an action spectrum for skin cancer can only be obtained from animal experiments. The albino mouse is the model generally used for studies of UV-induced skin cancer, notably SCC, and after taking into account differences in the optics of human epidermis and albino mouse epidermis, the experimentally determined action spectrum for tumour induction in mouse skin can be modified to arrive at a postulated action spectrum for human skin cancer. The resulting photocarcinogenesis action spectrum is published as an international Standard by the Commission Internationale de l'Eclairage [15]. This action spectrum is compared in figure 4.20 with that for erythema and given the similarity between the two action spectra, erythemally-weighted doses, often in units of SED, are sometimes used as a surrogate for carcinogenic-effective doses.

A defensible estimate of the action spectrum for melanoma remains elusive. Early work in a hybrid tropical fish suggested a role for UVA but this is not supported by later work in the same model or in mammalian models. In fact, the weight of evidence now seems to support UVB as critical to the initiation of melanoma, with perhaps a contributory role for UVA in the progression of the disease.

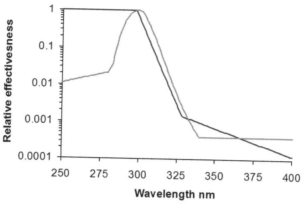

Figure 4.20. The action spectra for photocarcinogenesis (red) and erythema (blue).

4.4.9 Dose response

Estimates of the risk of inducing skin cancer from exposure to solar UV radiation demands knowledge of dose–response relationships. These data remain unknown for malignant melanoma and so it is unwise to make predictions about the relative risk of inducing melanoma from either elective or adventitious sun exposure. However, data on dose–response relationships are available to some extent to allow quantitative estimates to be made of the risk of inducing non-melanoma skin cancer from sun exposure.

The application of multivariate analysis to population-based epidemiology of NMSC has shown that, for a group of individuals with a given genetic susceptibility and age, environmental UV radiation exposure is the most important factor in determining the relative risk. This leads to a simple power-law relationship which expresses the risk of NMSC for a constant annual solar UV radiation exposure as:

$$\text{Risk} \propto [\text{Annual solar UV exposure}]^{\beta}$$

The exponent β is termed the biological amplification factor and is normally derived from surveys of skin cancer incidence and UV radiation climatology at different geographical locations. The exponent, β, is around 2.7 for SCC but lower at 1.4 for BCC [16].

This relationship is applicable to situations where the annual UV exposure received by an individual remains unaltered throughout life. In those instances where this assumption does not hold, for example, when cosmetic tanning equipment is used for a limited period during life, more complex models are required.

4.5 Photoageing

When we look at figure 4.21 (top), we see facial skin that is wrinkled and sagging with pigmentary changes, whereas the skin of the lower back in figure 4.21 (bottom) is smooth and elastic. We could reasonably conclude that the person in figure 4.21 (top) is much older than in figure 4.21 (bottom), but we would be wrong as they are one and the same person—a 68 year old female.

So whilst both areas of skin have the same chronological age, their exposure to solar radiation over the course of the subject's lifetime has been vastly different. The facial skin shown in figure 4.21 has been exposed to solar UV radiation on a daily basis, the magnitude of exposure varying considerably over the course of a year reflecting, among other things, seasonal variations in ambient UV radiation and time spent outdoors each day. On the other hand, the skin of the lower back in figure 4.21 has been protected from solar UV on almost all days throughout life and exposed only occasionally during outdoor recreational activities such as swimming. So whilst the skin of the lower back reflects the changes of chronological ageing, the

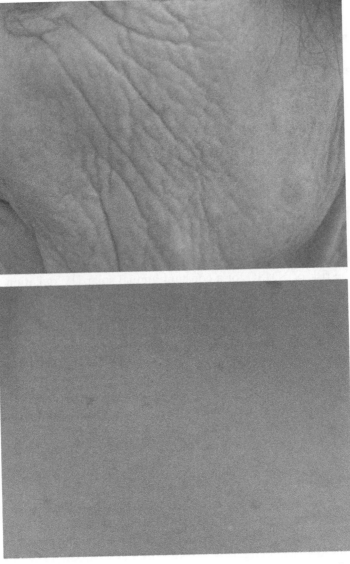

Figure 4.21. The face (top) and lower back (bottom) of a 68 year old female.

changes we see on the face are largely the result of chronic UV exposure that leads to premature skin ageing, referred to as *photoageing*.

Photoageing is characterised by fine and coarse wrinkling of the skin, mottled pigmentation, sallow colour, textural changes, loss of elasticity, sagging and premalignant actinic keratoses. Most of these clinical signs are caused by alterations in the dermis. Pigmentary disorders such as lentigines and diffuse hyperpigmentation are characteristic of epidermal changes.

Photoageing is quite distinct from chronological, or intrinsic, ageing in which the main clinical signs of ageing on sun-protected skin are dryness, fine wrinkles, skin atrophy, homogeneous pigmentation, and seborrheic keratoses.

Although all races are susceptible to photoageing, people with higher skin phototypes (IV–VI) are less susceptible to the deleterious effects of UV irradiation than people with a lower skin type. This phenomenon is most likely a result of the protective role of melanin.

The mechanisms of photoageing are thought to include the production of reactive oxygen species, which activate cell surface receptors, mitochondrial damage, protein oxidation and telomere-based DNA damage responses.

4.5.1 Action spectrum

The relative importance of different wavelengths in ageing human skin cannot be readily determined because of the long latent period and slow evolution of photoageing. Instead, extrapolation from experiments using hairless mice or the miniature pig is relied upon. Since approximately one half of UVA radiation and less than 10% of UVB radiation incident on white skin penetrates to the dermis, it is not surprising that results from animal studies have shown that chronic UVB and UVA irradiation in hairless mouse skin both result in histological, physical and visible changes characteristic of photoageing.

It should be remembered that solar radiation includes not only UV radiation but also visible and infrared radiation. Visible light is thought to play little role in photoageing but studies have shown that infrared (IR) radiation can induce cutaneous angiogenesis and inflammatory cellular infiltration, reactive oxygen species, disrupt the dermal extracellular matrix by inducing matrix metalloproteinase, and alter dermal structural proteins. These effects have been observed both in cellular models and in human skin *in vivo* and have been taken to infer that IR radiation may contribute to premature skin ageing. How important this contribution may be in the context of typical human exposure to sunlight remains controversial (see section 7.15).

4.6 UV effects on the eyes

Although this book is concerned principally about protecting the skin from excessive sun exposure, it should be remembered that the eyes can also be damaged by exposure to solar UV radiation. However, ocular exposure is far more affected by aversive behaviour than the skin. Although the cornea is more sensitive to UV injury than the skin, acute eye damage rarely occurs in situations where sunburn occurs, except possibly on snowfields. This is because we tend to shield our eyes from strong sunlight either by moving our eyes away from the Sun or by narrowing the eyelid aperture i.e. squinting.

The acute effects of UV exposure are primarily those of conjunctivitis and photokeratitis. Whilst acute eye damage is rare from solar UV exposure, it is a real risk from artificial sources of UV, especially those emitting high levels of UVC and UVB accompanied by weak levels of visible light.

Conjunctivitis is an inflammation of the membrane that lines the insides of the eyelids and covers the cornea, and may often be accompanied by an erythema of the skin around the eyelids. There is the sensation of 'gritty eyes' and also varying degrees of photophobia (aversion to light), lacrimation (tears), and blepharospasm (spasm of the eyelid muscles) may be present.

Photokeratitis is an inflammation of the cornea which can result in severe pain. Ordinary clinical photokeratitis is characterised by a period of latency that tends to vary inversely with the severity of UV exposure. The latent period may be as short as 30 min or as long as 24 h, but it is typically 6–12 h. The acute symptoms of visual incapacitation usually last from 6–24 h. Almost all discomfort usually disappears within two days and rarely does exposure result in permanent damage. Unlike the skin, the ocular system does not develop tolerance to repeated exposure to UV radiation.

Cases of photokeratitis have been reported following reflection of solar UV radiation from snow and sand but it is more commonly seen from industrial sources such as electric welding arcs when sufficient care has not been taken to protect the eyes. For this reason, the condition is sometimes referred to as 'snow blindness', 'arc eye' or 'welder's flash'.

Chronic effects attributed to solar UV exposure on the eyes include pterygium, ocular melanoma and cataract formation.

Pterygium is a wing-shaped overgrowth of the conjunctiva, which in extreme cases may obstruct vision. Epidemiological studies have implicated solar UV exposure as a causal factor.

There is limited evidence, and far less conclusive than it is for skin melanoma, that solar UV radiation is linked to the development of ocular melanoma.

Chronic exposure to sunlight over many decades may result in cataract formation, which is a partial or complete loss of transparency of the lens or its capsule. The mechanism of formation of cataracts probably involves several factors but from epidemiological studies there is some evidence that the UVB component of sunlight plays a role and may be more important than the UVA component.

References

[1] Fitzpatrick T B 1975 Soleil et peau *J. Med. Esthet.* **2** 33–4
[2] Harrison G I and Young A R 2002 Ultraviolet radiation-induced erythema in human skin *Methods* **28** 14–9
[3] McKinlay A F and Diffey B L 1987 A reference action spectrum for ultraviolet induced erythema in human skin *CIE J.* **6** 17–22

[4] Commission Internationale de l'Eclairage 1998 Erythema reference action spectrum and standard erythema dose CIE S007E-1998 (Vienna: CIE Central Bureau)

[5] Parrish J A, Jaenicke K F and Anderson R R 1982 Erythema and melanogenesis action spectra of normal human skin *Photochem. Photobiol.* **36** 187–91

[6] Sklar L R, Almutawa F, Lim H W and Hamzavi I 2013 Effects of ultraviolet radiation, visible light, and infrared radiation on erythema and pigmentation: a review *Photochem. Photobiol. Sci.* **12** 54–64

[7] Menagé H du P, Harrison G I, Potten C S, Young A R and Hawk J L M 1995 The action spectrum for induction of chronic actinic dermatitis is similar to that for sunburn inflammation *Photochem. Photobiol.* **62** 976–9

[8] Ervik M, Lam F, Ferlay J, Mery L, Soerjomataram I and Bray F 2016 *Cancer today* (Lyon: International Agency for Research on Cancer) Available at: http://gco.iarc.fr/today (last accessed 18 April 2017

[9] Ahmad O E, Boschi-Pinto C, Lopez A D, Murray C J L, Lozan R and Inoue M 2000 *Age standardization of rates: a new WHO standard (GPE Discussion paper Series: No. 31)* (Geneva: World Health Organization)

[10] Doll R 1971 The age distribution of cancer: implications for models of carcinogenesis *J. R. Stat. Soc.: Series A (General)* **134** 133–66

[11] US Department of Health and Human Services 2014 *The Surgeon General's call to action to prevent skin cancer* (Washington DC: US Department of Health and Human Services, Office of the Surgeon General)

[12] Subramaniam P, Olsen C M, Thompson B S, Whiteman D C and Neale R E 2007 Anatomical distributions of basal cell carcinoma and squamous cell carcinoma in a population-based study in Queensland, Australia *JAMA Dermatol.* **153** 175–82

[13] Surveillance Research Program of the Division of Cancer Control and Population Sciences, National Cancer Institute Available at: http://seer.cancer.gov/seerstat

[14] National Cancer Registry Available at: www.nioh.ac.za/?page=cancer_statistics&id=163

[15] Commission Internationale de l'Éclairage 2006 *Photocarcinogenesis action spectrum (nonmelanoma skin cancers) CIE S 019/E:2006* (Vienna: CIE Central Bureau)

[16] de Gruijl F and van der Leun J C 1993 Influence of ozone depletion on the incidence of skin cancer: quantitative prediction *Environmental Photobiology* ed A R Young, L O Björn, J Moan and W Nultsch (New York: Plenum Press) pp 89–112

Sun Protection
A risk management approach
Brian Diffey

Chapter 5

The impact of time and space in moderating human exposure to solar UV radiation

Before looking at ways of how we can separate humans from solar UV radiation, we need to understand the extent to which we are exposed to sunlight.

5.1 Behavioural influences on exposure to solar UV radiation

Daily ambient erythemal UV radiation shows a clear-sky summer-to-winter ratio of about 20:1 in temperate latitudes (~50°), falling to about 3:1 at latitudes around 30°, with day-to-day perturbations superimposed on this annual cyclic pattern as a result of cloud cover. However, the UV exposure of an individual living at a specific location will exhibit much greater fluctuations than ambient variation because of differences in time spent outdoors and proximity to shade on different days throughout the year. Furthermore, the UV dose absorbed by the skin is further modified by the use of photoprotective agents such as hats, clothing and sunscreens.

Broadly speaking, we can divide our sun exposure into *adventitious* exposure, typified by the unavoidable exposure associated with activities such as shopping and travelling to work (where exposed sites are normally limited to the face, neck and hands), and *elective* exposure, when we deliberately go to seek the sun for recreational purposes, usually during summer weekends and holidays. During our elective exposure we often expose our arms, legs and sometimes our trunk or even whole body.

During weekdays, it is likely that an indoor worker may be outside only in urban areas where nearby buildings will often obscure direct sunlight and a large part of the sky, as illustrated in figure 2.3(b). Under these conditions exposed sites may only receive about 5%–25% of the UV that is incident on an unshaded, horizontal surface. At weekends, especially during recreational exposure, more time may be spent away from urban areas where a much greater part of the sky will be visible and shade of direct sunlight less frequent.

5.1.1 Time spent outdoors

How long we spend outdoors is a major contributor to our annual UV burden. The majority of people in developed countries work mainly indoors for 5 days per week, resulting in about 105 weekend days per year. In addition, if we include the number of paid vacation days (including public holidays), which is typically 28 days, this gives something like 133 days per year when there is the opportunity to be outdoors. However, statutory paid leave varies considerably from zero days in the US, 10 days in Japan up to around 40 days in Austria.

There are large variations between countries in the number of days that children spend at school each year. For example, in Bolivia children are at school for 160 days per year, whereas in China the figure is 260 days. A more typical value for many countries is around 190 days, leaving 175 days per year when there is the opportunity to be outdoors.

Unlike adults, children tend to have longer lunch breaks and often spend part of the school day outside engaging in activities such as physical education. This, coupled with more days away from the 'workplace' means that children have much greater opportunity for sun exposure than adults.

There are two principal ways to estimate time per day spent outdoors: self-reported time using personal diaries or recorded by time-stamped personal dosimeters; and recall studies asking people to estimate the time they spend outdoors during defined periods.

Whilst there will be an appreciable variation in the time spent outdoors during a specific exposure period (e.g. summer weekdays) by habitués, generally studies have shown that times spent outdoors by a specific cohort, e.g. children attending a particular school, are positively skewed.

Figure 5.1 shows the results from an overview analysis [1] of the median times per day spent outdoors by adults from the USA and several European countries during weekdays and weekends as recorded in a number of different studies either by diaries, dosimeters, or by recall.

It is striking how heterogeneous these data are. Sources of heterogeneity are not confined to the differences in geographical location, duration, time period during the day and months of the year when sampling took place. There will also be differences in the attributes of the cohorts in each study that include not just age and gender, but other factors such as the physical activity level, health and employment status of individuals. Additionally, weather factors of daily maximum temperature and precipitation will also contribute to the variance.

Notwithstanding the heterogeneity that exists between the various studies, the pooled estimates of median times per day outdoors were around 1.0 h and 1.5 h for weekday and weekend exposure, respectively. The pattern of time outdoors for holiday exposure more closely followed a normal distribution with a mean time per day outdoors of about 5 h. Although these summary values should be regarded cautiously, they may still provide a better estimate of exemplary values for the population than using the data obtained in any single study (Stein's paradox in statistics).

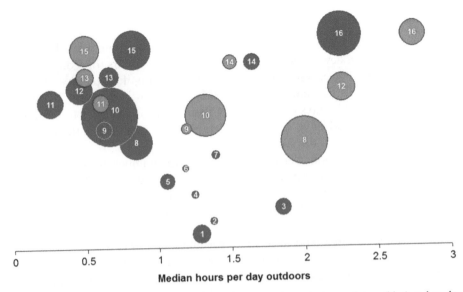

Figure 5.1. An overview of the median times per day spent outdoors during weekdays (blue) and weekends (red). The symbol area represents the sample size of individual studies, and the numbers in the circles are the identifiers of studies that contributed to the overview (adapted from reference [1]).

5.1.2 Measuring personal exposure to solar UV radiation

Estimates of personal UV exposure are normally obtained by direct measurement using either UV-sensitive film badges, most commonly polysulphone film (figure 3.11), or electronic dosimeters generally worn on the wrist (figure 3.10). The results obtained from a number of studies in mid-latitudes indicate broadly that people receive an exemplary annual exposure of the order of 200 SED mainly from summer weekend and vacational exposure, and principally to the hands, forearms and face. Excluding vacations to sunny places, adults who work outdoors receive about 10% of the total ambient available on a horizontal plane, while indoor workers and children generally receive around 2%–5% of ambient. Depending on the type of behaviour during summer vacations, a two week sunbathing holiday in a region of high insolation can contribute up to 100% of the cumulative UV exposure in the remaining 50 weeks of the year for an indoor worker.

On a population basis, however, annual exposure can vary enormously depending on an individual's propensity for being outdoors. For example, the annual solar UV doses received by 164 Danish volunteers (69 males and 95 females) within the age range 4–67 years were measured using both sun exposure behaviour recorded in diaries and by personal UV dosimeters that measured time-stamped UV doses [2]. There was a wide range of individual exposures within the cohort extending from a few tens of SED to several hundred SED, with a median annual UV dose of 166 SED for the total group, as shown in figure 5.2.

The authors found that the annual UV dose did not correlate with age, although the variation in annual UV dose was high. From a review of personal UV doses

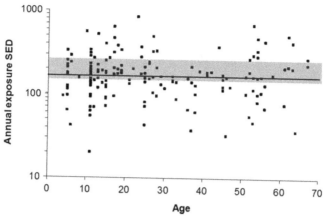

Figure 5.2. Annual UV doses of 164 children and adults in Denmark [2]. The horizontal line shows the median annual UV dose of 166 SED for the total group. The grey band indicates the range of modelled annual personal UV exposures (see section 5.4).

worldwide [3], it was estimated that cumulative UV exposure increases approximately linearly throughout life. However, with the trend for earlier retirement, greater life expectancy and higher disposable income with opportunities for overseas travel, we may see this fraction diminish as the sun exposure in late middle age and onwards contributes a larger fraction than hitherto of lifetime UV exposure.

5.1.3 Impact of behaviour on personal UV exposure

Even within an individual their solar UV exposure can show wide variations over a short time interval. This is illustrated in figure 5.3, which shows the results of a subject wearing an electronic dosimeter pinned to the chest and wandering around an open air market around noon in the summer.

The erratic nature of personal UV dose rate in just this 20 min period reflects behaviours such as turning toward and away from the sun, stooping forward, and moving into the shade of the awnings of the market stalls.

On a population basis, people living in sunny countries receive, on average, higher UV exposure than those who live in more temperate latitudes. However, on an individual basis there are large variations in daily UV exposures within a cohort of subjects. This is illustrated in figure 5.4, which shows frequency histograms of daily outdoor UV exposures received in the summer months by schoolchildren in England and Queensland, Australia [4], regions with large differences in ambient UV radiation.

Whilst the median daily personal exposure in Queensland (1.7 SED) was twice that received in England (0.9 SED), there was a wide overlap between the two distributions. On average, children in both countries received about 5% of ambient, yet the distribution of exposures in Queensland was considerably greater than in England. Children in England rarely received a daily exposure above 3 SED, an approximate threshold above which clinical signs of erythema might be seen. In the Queensland children, however, this exposure was exceeded on one third of occasions and so whilst the median doses differed by a factor of two, roughly in proportion to

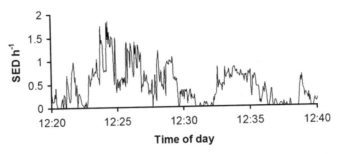

Figure 5.3. The impact of behaviour on personal sun exposure minute-to-minute.

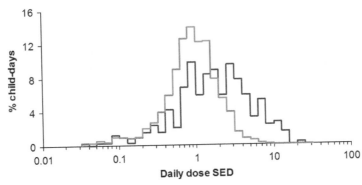

Figure 5.4. Frequency histograms of daily outdoor UV exposures received in the summer by schoolchildren in England (red) and Queensland (blue) (data from reference [4]).

differences in ambient, the incidence of sunburn in Queensland would be expected to be considerably more common than in England, assuming no photoprotection interventions were in place.

The significant influence that behaviour can have on individual sun exposure is exemplified by the finding that on any one summer day we might expect the daily solar UV exposure of 17% English children to exceed the Queensland median, and for the exposure of 26% of Queensland children to be less than the English median.

Another finding from this study was that children did not behave as a homogeneous group with respect to outdoor exposure in that for both weekday and weekend exposure some children, week-on-week, received consistently higher or lower exposures in relation to their peers.

5.1.4 Anatomical distribution of sunlight

So far we have been assigning a single value to a given individual's UV exposure. It is evident, however, that the human form, by virtue of its complex topology, receives highly variable solar exposure across its surface area. A number of both measurement and modelling studies have looked at UV exposures, relative to ambient, at different anatomical sites.

Whilst a number of climatic, postural and behavioural factors will influence the exposure at a given anatomical sites, as illustrated in figure 5.3, as a rule of thumb, we can say that an ambulant subject in mid-latitudes moving randomly under an open sky between mid-morning and mid-afternoon during the summer season (solar altitudes typically 45° to 65°) receives an exemplary exposure to the face and trunk of one-third of that of ambient on a horizontal surface. The legs receive a little less at about one-quarter of ambient, whereas the epaulet regions of shoulders receive about two-thirds of ambient.

At lower solar altitudes around 25°—early morning and late afternoon in the summer—this fraction rises to about two-thirds of ambient for vertical surfaces that include the face, trunk and legs, but at these times of day the UV index is around 1 and so there is little risk.

These values apply to an ambulant subject with no shade nearby. When the influence of shade is factored in, a person walking around in an urban environment, as illustrated in figure 2.3b, would receive an exposure on vertical body surfaces (hands, face, arms, legs) of typically less than one-sixth of the ambient exposure due to the combined effect of body geometry, random orientation with respect to the sun and partial obstruction of the sky (and possibly direct shading of sunlight) by nearby buildings.

5.1.5 Ambient temperature and personal UV exposure

Ambient temperature influences the magnitude of individual solar UV exposure and the consequent risk of sunburn. In a behavioural study in Australia [5], it was observed that the likelihood of sunburn approximately doubled when the ambient temperature was in the range 19–27 °C, compared to temperatures of 18 °C or lower. The reason for this is presumably because warmer temperatures encourage people to spend more time in direct sunlight with the increased risk of sunburn. Interestingly, the study found that at temperatures in excess of 27 °C, the likelihood of sunburn fell again as people sought shade for comfort reasons.

Greenhouse-gas induced climate change is expected to change ambient temperatures, which could impact on people's behaviour and the time they spend outdoors. A change to warmer conditions is predicted to occur in many countries, where in the UK, for example, the average annual temperatures may rise by between 2 °C and 3.5 °C by the 2080s. These average temperature changes will be accompanied by an increased frequency of extreme temperature events and high summer temperatures, as well as wetter winters and drier summers. Clearly such changes in climate could encourage behaviour that may increase, or decrease, population exposure to sunlight and the health risks associated with it.

5.2 Attitudes to sun exposure

Whilst climatic factors may influence the levels of UV radiation at the Earth's surface, it is the behaviour of people outside that has a much greater impact on personal UV exposure, and so a combination of environmental, socioeconomic and behavioural factors influence an individual's exposure to solar UV radiation. It goes without saying that most people are stimulated on a blue sky, sunny day and are maybe not quite so cheerful on a rainy, overcast day. People living in more northerly latitudes feel starved of sunshine and warmth for so much of the year that when the opportunity does arise to spend time in the sun, either at home or abroad, most are only too pleased to take advantage of this.

Whilst the pleasure of simply being outdoors is shared by people of all ages, a noticeable change takes place during teenage years when people realise the potential of the sun to cause tanning and so to enhance their appearance. Although there is widespread public awareness of the dangers of overexposure, sunbathing remains popular and there is a widespread feeling that 'sunlight is good for you'. Whist the physiological effects that presumably underlie the feeling of well-being remain to be explained adequately, there is little doubt that sun exposure has beneficial effects, as discussed in section 2.3.

However, attitudes are changing. Since 2003, Cancer Research UK has commissioned the Office of National Statistics to run a *SunSmart* tracking survey aimed at monitoring trends in the general public's approach towards sun protection. A recent report compared awareness and behavioural trends between 2003 and 2013 [6]. In terms of behaviour, the results showed significant positive trends observed for 10 out of 14 behaviours focussed on avoiding the sun or skin cancer, which included a greater percentage of people now using SPF15+ sunscreen, wearing a hat, reducing time in the sun, avoiding sunburn, and avoiding sunbeds.

5.3 Trends in sun exposure

Lifestyles are changing in a way that impact on personal exposure to the sun and the use of photoprotective measures, nowhere more so than the opportunity for overseas travel. The total number of visits abroad by UK residents in 2015 was 65.7 million and of these 42.2 million were for holidays [7].

Holiday visits, which account for nearly two thirds of visits abroad by UK residents, have increased dramatically with a ten-fold increase in the number of overseas holidays taken by British residents in the period 1971–2015. The number of holidays overseas peaked at just over 45 million in 2008, dropped by 16% the following year possibly as a consequence of the economic downturn, but are now on the rise again (figure 5.5).

Furthermore, holiday travel by British people, and probably by people from other northern European countries, is overwhelmingly to more southerly destinations where UV levels are typically high. For example, Spain continued to be the top destination for UK residents visiting abroad in 2015, accounting for 13.0 million visits, a 6.1% increase from the previous year, and accounting for 20% of the total number of visits abroad.

Clearly changing patterns of holidaymaking are an important factor that has increased the overall UV burden received by the many populations from more northerly latitudes in recent decades.

As well as holidays to sunny destinations, outdoor pursuits remain popular, with consequential sunlight exposure. For example, in 2012 nearly half of all Americans participated in outdoor activities, with the most popular activities being running/jogging, angling, cycling and hiking [8].

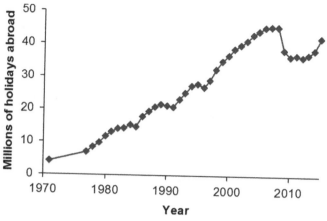

Figure 5.5. Holidays abroad taken by UK residents in the period 1971–2015 [7].

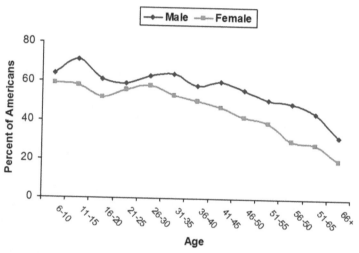

Figure 5.6. The percentage of American people by age and gender who engage in outdoor recreation pursuits [8].

Participation in outdoor recreation varies as individuals age, with gender also playing a role in determining behaviours and participation trends (figure 5.6).

Outdoor activities are popular among children, especially among boys ages 11 to 15. Participation rates drop for both males and females from ages 16 to 20 and then climb back up slightly for females in their early 20s and males late 20s before gradually declining throughout life. Participants in outdoor recreation represent a diverse population by age, ethnicity, income, and education.

5.4 Modelling human exposure to solar UV radiation

Measurement of personal UV exposure demands high compliance over an extended period of time by a large number of people. Furthermore, there is a great deal of heterogeneity in published studies concerning factors such as numbers of subjects monitored, geographic location, period of study (ranging from a few days to sampling different periods throughout the year), anatomic site of dosimeter placement and data presentation.

An alternative approach is to model the variables that affect personal exposure by combining ambient UV exposure with the exposure on critical sites (e.g. face) relative to ambient UV and time spent outdoors. When we look at figure 5.3, we realise how erratic individual UV exposure is even over a short time period and so the reliability of modelling might be questioned. However, on a population basis these random fluctuations in time are averaged out somewhat and so it is possible to derive representative estimates of 'typical' population exposure.

Whilst there is no limit to the complexity of modelling human behaviour outdoors, by way of illustration we adopt the principle of Ockham's razor and describe a very simple model that estimates a representative average annual personal solar UV exposure. We express this as:

$$\text{Annual UV exposure} = \text{UV}_{ambient} \times [5 \times EF_{wd} \times f(h_{wd}) + 2 \times EF_{we} \times f(h_{we})]/7$$

$\text{UV}_{ambient}$ is the typical ambient annual erythemal UV at the location of choice.

EF is the fraction of ambient UV incident on the anatomical region of interest; we consider the face as this site receives constant sun exposure whenever outdoors and is the region most likely to exhibit photoaging or develop skin cancer. We assume from section 5.1.4 that during weekdays EF_{wd} is one-sixth of ambient (mostly urban exposure), and that during weekends EF_{we} is one-quarter of ambient (mixed urban and non-urban exposure).

h are the hours spent outside each day; we assume that during weekdays h_{wd} is 1.0 h, and that during weekends h_{we} is 1.5 h (see section 5.1.1).

In order to maintain a simple approach, it is assumed that the hours spent outside on any given day are symmetrical about solar noon and that the irradiance of solar erythemal UV exhibits a triangular distribution between sunrise and sunset, peaking at solar noon, the latter assumption being well supported from the diurnal variation of erythemal UV at different latitudes. The fraction, $f(h)$, of daily ambient UV available is then given as $1 - [1 - h/H]^2$, where H is the mean hours of daylight averaged over the year ($H = 12$ h for all latitudes). So we estimate f_{wd} and f_{we} to be approximately one-sixth and one-quarter, respectively.

If we run this model for Denmark (latitudes 55–57 °N), we estimate from figure 3.14, that a typical ambient annual exposure is ~4000 SED. Consequently, we calculate:

Annual facial exposure on weekdays: $4000 \times [5 \times 1/6 \times 1/6]/7 \sim$ **80 SED**

Annual facial exposure on weekends: $4000 \times [2 \times 1/4 \times 1/4]/7 \sim$ **70 SED**

Total Annual Exposure ~ **150 SED**

We can use a similar approach to estimating holiday exposure, for example a 2 week summer beach holiday in southern Europe, as follows:

Typical ambient UV on a summer day in southern Europe taking into account cloud cover	45 SED
Time outdoors per day on holiday	Typically 5 h
Fraction of daily ambient UV for 5 h exposure	About one-half
Fraction of ambient received assuming mixture of time spent ambulant and sunbathing	About one-third
Daily UV exposure on skin	$45 \times \frac{1}{2} \times \frac{1}{3} \sim 7.5$ **SED**
Two-week holiday exposure	$14 \times 7.5 \sim 100$ **SED**

We can summarise the range of exemplary annual personal solar UV exposures of northern Europeans as about 150 to 250 SED, depending on propensity for spending a summer holiday in the sun. This range of average exposures is shown by the gray shaded region in figure 5.2 where we can see that the modelled exposures encompass the measured median annual exposure of these 164 individuals. But what is also clear is that these estimates are just representative of a mean population exposure and as figure 5.2 shows, annual individual exposures will show much greater variation depending on propensity for being outside and the types of activities and behaviours associated with this.

5.5 Strategies for controlling human exposure to solar UV radiation

5.5.1 When sun protection is needed

Figure 5.7 is a sketch summarising the approximate contributions to the annual solar UV radiation exposure to people living in mid-latitudes. We see that the single largest contribution—about three-quarters of annual exposure—is from exposure during the summer (May to August), including a two-week summer holiday to a sunny destination.

Consequently, if we wish to limit, or control, our annual solar UV burden, we need to focus on recreational and vacational exposure during the summer. Over-concern

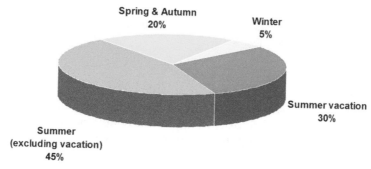

Figure 5.7. The approximate contributions to annual solar UV burden of people living in mid-latitudes (Winter: November to February; Spring: March and April; Summer: May to August; Autumn: September and October).

about our exposure during the six-month period October through to March (at temperate latitudes in the northern hemisphere) will have little impact on our overall annual exposure and could be detrimental in terms of our vitamin D status.

5.5.2 Education

The most productive way to inform the public about measures to control exposure to the sun is through education. This can come from a variety of sources that include government, meteorological services, cancer organisations, healthcare professionals, the media, and schools.

Government
Governments can contribute to controlling their population burden to UV radiation either by legislation or, more commonly, by disseminating information through appropriate channels.
　　Just two examples of legislative control are:
- Some states in the US have passed legislation to support sun-safety education programmes incorporating awareness about skin cancer prevention in both schools and to state employees who work outdoors.
- A contribution to population UV exposure, in addition to solar UV radiation, results from the use of sunbeds for cosmetic tanning. Because of the rising concern about the detrimental effects of sunbed use, many countries have enacted legislation to tighten regulations on the sunbed industry. A total ban is in place in Brazil, and legislation prohibits use by people under 18 years in the UK, several other European countries, Australia, and parts of Canada and the US.

　　Many countries can demonstrate governmental leads in skin cancer and sun protection education but as just one example the UK Government, through its agency the *National Institute for Health and Care Excellence*, published recommendations in 2016 for health and social care practitioners to communicate consistent, balanced messages about the risks and benefits of sunlight exposure [9]. And in the spring of 2013, the UK Government agency *Public Health England* raised a sunburn warning after measurements of ambient UV radiation at sites across the UK revealed unusually high levels.

Meteorological services
These services can provide a vital role in providing the public with information on the expected UV index (see section 3.2.2) for the current and forthcoming days. In estimating the UV index, forecasters take into account the solar altitude, ozone amounts in the stratosphere and the expected cloud cover.

Cancer organisations and healthcare professionals
Skin cancer prevention activities were first initiated in Australia from the 1960s, and later, coordinated primary prevention campaigns from the early 1980s (*Slip! Slop!*

Slap!) were introduced by the State Cancer Councils with the objective to encourage individuals to reduce their exposure to UV radiation. In 1988, the *Slip! Slop! Slap!* campaign was superseded by the more comprehensive *SunSmart* skin cancer prevention programme, which, amongst other things, provides many resources such as the poster illustrated in figure 5.8.

Beginning in 1993, the *Sun Know How* campaign was developed by the (now defunct) Health Education Authority in the UK with the aim of raising public awareness about the health consequences of overexposure to the sun. This campaign was replaced in 2003 by *SunSmart*, the UK's national skin cancer prevention campaign, hosted until 2012 by Cancer Research UK with funding from the UK health departments. The *SunSmart* campaign incorporates research, public communication, working with professionals and policy development to address the burden of skin cancer in the UK, and drew on the experiences of the Australian and other national and international health programmes.

In addition, other organisations in the UK promote similar sun awareness activity; for example, the British Association of Dermatologists, which runs a national campaign called *Sun Awareness* that includes national Sun Awareness Week each May, and *Skcin: the Karen Clifford skin cancer charity*.

In the US, the Skin Cancer Foundation and the American Academy of Dermatology are major bodies active in sun awareness and skin cancer prevention, and many other countries can demonstrate similar activity by their health agencies.

The media

Traditional media, such as magazines, newspaper, radio, and television, play an important role in disseminating information related to sun protection, especially the role of sunscreen, and it is through these channels that most people learn about UV effects on the skin and photoprotection.

More recently, the internet and social media have supplemented the messages from traditional media with the availability of websites and apps related to, for example, sun safety, the UV index and when to re-apply sunscreen. Some of these apps use satellite data to determine an individual's location, and use these data to estimate the UV index followed by giving personal sun exposure advice based on solar UV intensity and the individual's skin type, determined in response to answers to questions requested by the app.

Schools

School-based educational programmes are intended to teach children at a young age about the impact of sun exposure on their skin health and steps that can be taken to minimise exposure, with the hope that this may foster lifelong habits regarding sun protection.

One such programme in the US is the *SunWise School Program* initiated by the US Environmental Protection Agency. In the Australian state of Victoria, a *SunSmart* accreditation programme has operated in primary schools since 1994, and has now been adopted throughout Australia. Accredited schools adopt a sun protection policy that includes compulsory hat wearing for children playing outside

Figure 5.8. A poster of Sid Seagull surrounded by the 5 SunSmart steps: Slip, Slop, Slap, Seek and Slide (image published with kind permission of Cancer Council Victoria, 2017).

in the summer terms, provision of adequate shade structures, and including sun protection education in the curriculum.

5.5.3 Community interventions

Alongside education, community interventions in controlling UV exposure can be helpful, especially in the provision of well-designed shade structures (see chapter 6).

An important contributor to community interventions is the Community Preventive Services Task Force, which was established in 1996 by the US Department of Health and Human Services. Its role is to identify population health interventions that are scientifically proven to save lives, increase life spans, and improve quality of life. The Task Force produces recommendations to help inform the decision making of federal, state, and local health departments, other government agencies, communities, healthcare providers, employers, schools and research organisations.

Some of the interventions recommended by the Task Force in the context of skin cancer risk and sun exposure relate to outdoor recreational and tourism settings, outdoor occupational settings, and child-care centres, schools and colleges.

One location where large numbers of people gather outside is at sporting events and appropriate scheduling by holding events, if possible, during early morning or late afternoon/evening can result in a reduced UV burden not only to the viewing public but also to the event participants. Other considerations in planning an outdoor event include ensuring there is adequate existing shade and/or providing temporary shade structures, and reminding spectators through event advertising, websites and social media to bring along appropriate personal sun protection, especially hats and sunscreen.

The major reason organisers consider the environment when scheduling outdoor sporting activities is the impact that high temperatures, sometimes coupled with high humidity, can have on the performance and health of the participants. The apparent temperature, sometimes called the *heat index*, (see table 5.1) indicates how hot it feels and is expressed as a function of ambient air temperature and the relative humidity.

When the temperature is high but the relative humidity is low, the heat index is less than the actual temperature. This is because cooling by evaporation of sweat is very efficient in these situations. However, when the relative humidity is high, evaporation is prevented and so it seems hotter than it really is. A combination of high temperature and high humidity leads to extreme heat indices and in these situations, physical activity outside can lead to heatstroke or even death.

Ambient temperature is often, but not always, positively associated with ambient UV levels and so by organising events to minimise heat load on participants, both the sportsmen and viewing public are likely to benefit from lower UV exposure.

One example of re-scheduling sporting events is the 2022 FIFA World Cup due to be held in Qatar. Traditionally this event is held in June and July but because of high summer temperatures that typically average around 42 °C, the event will be played in November and December when typical average temperatures are about 24 °C. The maximum UV Index in Qatar in the summer is typically 11 or higher but at the time the competition is scheduled for, it will be around 6.

5.5.4 Individual contribution to controlling UV exposure

Whilst external agencies can educate and guide our behaviour in the sun, the most significant contribution to an individual's annual solar UV burden is personal choice.

Shadow rule and UV index
A basic technique that can be used as a guide to the need for sun protection is to apply the so-called *shadow rule*. Put simply, if your shadow is longer than your

Table 5.1. Apparent temperature (heat index) in degrees Celsius according to air temperature and relative humidity [10].

Relative Humidity %	Air temperature °C										
	21	24	27	29	32	35	38	41	43	46	49
0	18	21	23	26	28	31	33	35	37	39	42
10	18	21	24	27	29	32	35	38	41	44	47
20	19	22	25	28	31	34	37	41	44	49	54
30	19	23	26	29	32	36	40	45	51	57	64
40	20	23	26	30	34	38	43	51	58	66	
50	21	24	27	31	36	42	49	57	66		
60	21	24	28	32	38	46	56	65			
70	21	25	29	34	41	51	62				
80	22	26	30	36	45	58					
90	22	26	31	39	50						
100	22	27	33	42							

Serious risk to health - heatstroke imminent

Prolonged exposure and activity could lead to heatstroke

Prolonged exposure and activity may lead to fatigue

height (typically in the early morning and late afternoon), your UV exposure is likely to be low and sun protection is generally not required. However, if your shadow is shorter than you are (typically around the middle of the day), the levels of UV radiation are sufficiently high that you would be advised to seek shade and protect your skin and eyes. Of course, the shadow rule only works if the sun is not covered by cloud.

The rationale for this rule is that solar altitude is the main determinant of solar UV intensity, which is generally expressed by the UV index. The maximum UV index, which occurs at solar noon, determined for different cloudless days throughout the year and at latitudes ranging from 0 °N to 60 °N [11] is plotted against the calculated solar altitude for the corresponding day and latitude in figure 5.9.

We see that UV index increases as the sun rises higher in the sky. There is variation in UV index around a given solar altitude due to temporal and geographic variations in ozone thickness. However, the relationship can be roughly expressed as:

$$\text{UV Index} = 14 \times \exp\{ -[(90 - A)^2/1800]\}$$

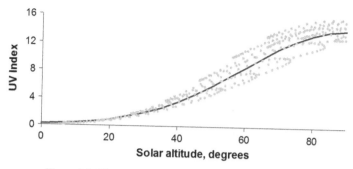

Figure 5.9. The variation of UV Index with solar altitude.

Table 5.2. The risk of sunburn and skin damage as a function of UV index for different skin types, together with the approximate range of solar altitudes and shadow lengths corresponding to the UV indices.

UV Index	Solar altitude	Shadow length relative to height	Sun-reactive skin type		
			I/II	III/IV	V/VI
0	<20°	>2.7	Very low	Very Low	Very Low
1–2	20°–35°	2.7–1.4	Low	Low	Low
3–4	35°–45°	1.4–1.0	Medium	Low	Low
5–6	45°–55°	1.0–0.7	High	Medium	Low
7–10	55°–65°	0.7–0.5	Very high	High	Medium
11+	>65°	<0.5	Extreme	Very high	High

where A is the solar altitude in degrees. This expression is shown by the solid curve in figure 5.9. From this expression we estimate that when your shadow length equals your height, that is a solar altitude of 45°, the UV index is typically around 4.5.

Interpretation of the UV index is dependent on an individual's propensity to sunburn, as illustrated in table 5.2.

When the risk is low, little needs to be done in the way of sun protection but as the risk rises then increasing vigilance and protection measures are needed. When shadow length is one-half or less of a person's height, UV levels are very high and considerable care needs to be taken.

Sun exposure at the coast

Many people believe that they sunburn easily at the coast because of reflectance of sunlight from the ocean. However, we see from figure 3.4 that reflectance of solar UV is generally between 15%–20% and so it seems that the reason people attribute getting sunburnt at the seaside is more to do with the absence of shade and reflection of UV from beach sand, rather than reflection from the sea. And since UV rays pass easily through water, swimming in either the sea or open-air pools offers little protection against sunburn.

Sunburn and ambient temperature

On days during the summer months when high temperatures are expected, weather forecasters frequently warn about the dangers of UV and high UV indices. It is not surprising, therefore, that it is a common belief that high ambient air temperatures are a major risk factor for burning. Although the UV index is generally higher on cloudless, hot days compared with cloudy, cool days, reliance should not be placed on ambient temperature alone as a guide to the need for sun protection.

This is illustrated in figure 5.10, which shows the UV index and air temperature in hourly intervals at Abuja, Nigeria (latitude 9.1 °N, longitude 7.4 °E, 840 m asl) on 16 March 2017, which was a cloudless day and where the solar altitude reached 80 degrees and solar noon occurred at the local time of 12:39 pm.

We see that in the morning both the air temperature and UV index increased steadily but in the afternoon the temperature continued to rise until mid-afternoon, whereas the UV index fell steadily. By 17 h, the UV index was only 1 and presented no risk of sunburn yet the air temperature was still in the mid-30s.

It is not uncommon to see people apply sunscreen in the late afternoon/early evening when it may still be very hot. However, in mid-summer the UV index four hours after solar noon is between 2 and 3 and one hour later, it is between 1 and 2 at latitudes between the Arctic and Antarctic Circles. At these UV indices, there is little risk of sunburn.

Consequently, it is unwise to judge the erythemal power of the Sun solely on ambient temperature. This is true not only when it is very hot but also when it is cloudy as the risk of overexposure may be increased because the warning sensation of heat is diminished.

Estimating daily ambient erythemal UV exposure from the UV index on clear days

Figure 3.12 indicates that solar UV irradiance on a cloudless day roughly follows a normal distribution between sunrise and sunset, peaking at solar noon, the time of maximum UV index (UVI_{noon}). Hence, the daily ambient erythemal UV (in SED) can be approximately calculated as:

$$\{ UVI_{noon}/40 \times [3600 \times H/5]\sqrt{2\pi}\}/100$$

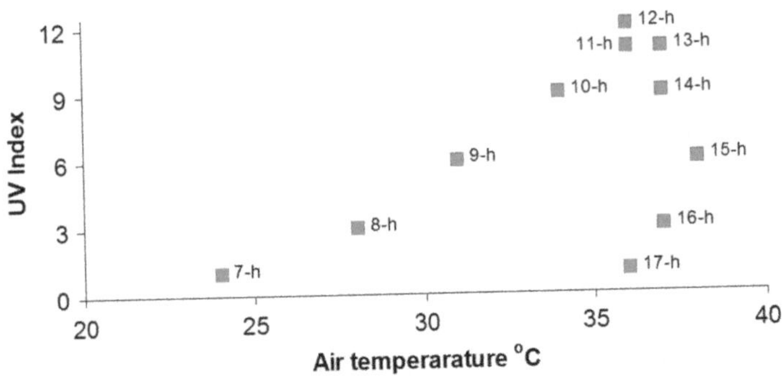

Figure 5.10. The air temperature and UV index in hourly intervals at Abuja, Nigeria on 16 March 2017.

A reasonable assumption is that the daylength (H) extends over about 2.5 standard deviations of the diurnal variation of irradiance, which accounts for the factor of 5 in the denominator of the second term in the expression.

This expression reduces to $0.45 \times \text{UVI}_{noon} \times H$, where H is the hours of daylight between sunrise and sunset for the date and latitude of interest (table 5.3).

As an example, the maximum daily ambient erythemal UV calculated using this simple approach for mid-June in the northern hemisphere is given in table 5.4.

When the sky is cloudy, this simple approximation is no longer valid as the variation in irradiance over the course of the day becomes irregular.

Sun avoidance around the middle of the day

Solar UV intensity is highest in the 3 h period around local noon when 40%–50% of a summer's day UV is received over the latitude range 10 °N to 50 °N. At more northerly latitudes, the long daylength in the summer means that only 25% of diurnal erythemal UV is received in this 3 h period at 60 °N, for example.

Avoiding direct sunlight around the middle of the day can be an effective first step in limiting personal exposure. This is illustrated in figure 5.11, which shows the cumulative ambient UV throughout a cloudless summer day in Tenerife (28°N) and Edinburgh (56°N).

Table 5.3. Daylength in hours for the mid-point of each month at latitudes in the northern hemisphere.

Month	0 °N	10 °N	20 °N	30 °N	40 °N	50 °N	60 °N
Jan	12	11.5	10.9	10.3	9.5	8.3	6.4
Feb	12	11.7	11.3	10.9	10.4	9.7	8.7
Mar	12	11.9	11.9	11.8	11.7	11.6	11.4
Apr	12	12.2	12.5	12.7	13.1	13.5	14.2
May	12	12.5	13.0	13.5	14.2	15.2	16.8
Jun	12	12.6	13.2	13.9	14.8	16.1	18.4
Jul	12	12.5	13.1	13.8	14.6	15.7	17.7
Aug	12	12.3	12.7	13.1	13.6	14.3	15.3
Sep	12	12.1	12.1	12.2	12.3	12.4	12.6
Oct	12	11.8	11.5	11.3	10.9	10.4	9.7
Nov	12	11.5	11.0	10.5	9.8	8.8	7.1
Dec	12	11.4	10.8	10.1	9.2	7.9	5.6

Table 5.4. Approximate maximum daily ambient erythemal UV for mid-summer in the northern hemisphere calculated from noontime UV index.

Latitude	UVI$_{noon}$	Daylength H hours	Daily ambient SED
20 °N	12	13.2	71
30 °N	11	13.9	69
40 °N	9	14.8	60
50 °N	7	16.1	51
60 °N	5	18.4	41

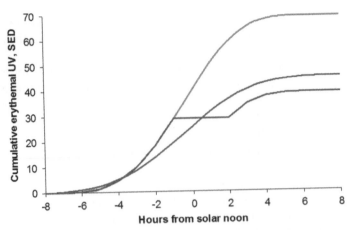

Figure 5.11. Cumulative ambient erythemal UV throughout a day in Tenerife (red curve) and Edinburgh (blue curve). Avoiding the sun for three hours around the middle of the day in Tenerife is shown by the green curve.

By the end of the day the cumulative ambient UV in Tenerife is 69 SED compared with 45 SED in Edinburgh. But if a leisurely 3 h lunch break is taken when on holiday in Tenerife, the cumulative UV at the end of the day, shown by the green curve in figure 5.11, is less than received during a full day in Edinburgh.

Holidays

As we have seen in figure 5.7, a summer holiday to a sunny destination can contribute a substantial fraction of our annual UV burden and one way to reduce this is to go on holiday (in the northern hemisphere) between October to March. As an example, the maximum UV index in Tenerife in June/July is 12 and daytime temperatures often reach 30 °C. In January, the weather is still pleasantly warm with average daytime temperatures between 16–21 °C but the maximum UV index is substantially lower at around 5.

Other options include winter ski resorts, or choosing summer holiday destinations in temperate rather than tropical latitudes, and/or opting for sightseeing holidays in preference to beach holidays and taking advantage of the shade provided by churches, galleries, museums and other buildings.

In many instances sun avoidance is not practicable as the sole means of controlling UV exposure and so it may be necessary to adopt personal protection in the form of shade, clothing and sunscreen, which form the subject of the following two chapters.

References

[1] Diffey B L 2011 An overview analysis of the time people spend outdoors *Br. J. Dermatol.* **164** 848–54

[2] Thieden E, Philipsen P A, Sandby-Moller J, Heydenreich J and Wulf H C 2004 Proportion of lifetime UV dose received by children, teenagers and adults based on time-stamped personal dosimetry *J. Invest. Dermatol.* **123** 1147–50

[3] Godar D E 2005 UV doses worldwide *Photochem. Photobiol.* **81** 736–49

[4] Diffey B and Gies P 1998 Time spent outdoors can be more important than ambient UV at an individual level *The Lancet* **351** 1101

[5] Hill D and Boulter J 1996 Sun protection behaviour: determinants and trends *Cancer Forum* **20** 204–10

[6] Cancer Research UK 2014 Trends in awareness and behaviour relating to UV and sun protection: 2003–2013 Available at: www.cancerresearchuk.org/sites/default/files/sun_protection_trends_-_cruk.pdf (last accessed 7 November 2003)

[7] ONS 2016 Travel trends: 2015 (London: Office for National Statistics)

[8] Outdoor Recreation Participation 2013 Boulder, CO: The Outdoor Foundation) Available at: www.outdoorfoundation.org/research.participation.2013.html (last accessed 21 April 2017)

[9] NICE 2016 sunlight exposure: risks and benefits (NG34) (London: National Institute for Health and Care Excellence)

[10] Steadman R G 1979 The assessment of sultriness. Part I: a temperature–humidity index based on human physiology and clothing science *J. Appl. Metereol* **18** 861–73

[11] Wester U and Josefsson W 1998 UV-index and influence of action spectrum and surface inclination WMO Global Atmosphere Watch GAW-report No **127** 63–6

IOP Publishing

Sun Protection
A risk management approach
Brian Diffey

Chapter 6

Physical barriers to protect humans from solar UV radiation exposure

The physical barriers that offer protection against solar UV radiation are shade, clothing and optical filters.

6.1 Shade

The solar UV radiation at ground level comprises a direct component from the sun and a scattered, or diffuse, component from the sky. When we seek shade and are in shadow, we remove the direct component and attenuate the diffuse component to a degree depending upon the fraction of the sky that is blocked by the shade structure (figure 6.1).

The purpose of shade is to protect people from excessive solar UV exposure that might be harmful to health, whilst at the same time to create an environment where people can enjoy the attributes of being outdoors—fresh air, warmth, and breeze.

Shade can be provided naturally by trees, by utilising canopies and semi-permanent structures, or by constructed shade in areas where large numbers of people may gather, such as sports stadia. An urban environment can provide substantial protection from diffuse solar UV radiation as shop awnings and tall buildings can reduce significantly the view of the sky (see figure 2.3(b)), and hence UV irradiance.

6.1.1 Shade as a neutral density filter

By taking the logarithm of the ratio of spectral irradiance of solar UV on a horizontal plane in full sun to that in shade, we can compute the spectral absorbance provided by the shade structure. This has been calculated assuming 60% occlusion of the sky, together with complete shading of direct sunlight, for an ozone thickness of 3 mm, and for solar altitudes ranging from 40° to 90°, with the results shown in Figure 6.2.

doi:10.1088/978-0-7503-1377-3ch6

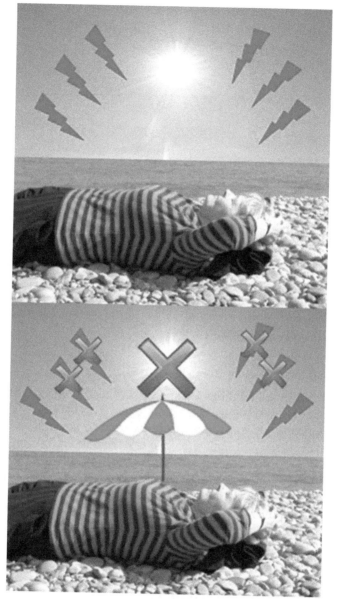

Figure 6.1. Top: when we are outside, we receive UV both directly from the sun and scattered from the sky; bottom: shading our skin from both direct sunlight and part of the sky will reduce UV intensity on our skin.

Also shown in figure 6.2 is the spectral absorbance determined from measured values of spectral irradiance on a horizontal plane in the shade of a tree at Toowoomba, Queensland (latitude 27.5°S), compared with the spectral irradiance in full sun at a distance of 3 m from the tree [1]. The measurement made in shade was in the shadow side of the trunk, in the approximate centre of the canopy shadow and at least 2 m from the visible shadow edge. The solar altitude at the

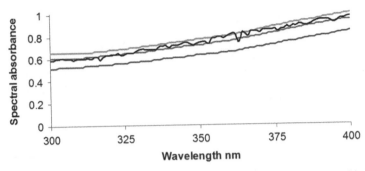

Figure 6.2. The effective spectral absorbance provided by shade for solar altitudes of 40° (blue), 60° (green) and 90° (red) assuming 60% occlusion of the sky, together with that measured under a tree (black) offering a similar level of shade.

time of measurement was 65° and approximately half of the sky was occluded beneath the tree.

It is clear from the approximate flatness of these curves that when we seek shade, nature is providing us with an approximate neutral density filter that reduces the intensity of solar UV on our skin with minimal change to the quality, or spectral mix, of UV. The effect of different amounts of shade is simply to shift the curves up and down without changing their shape.

6.1.2 Estimating the protection from shade

By weighting the diffuse and direct components of terrestrial spectral irradiance by the erythema action spectrum, we can estimate that over the range of solar altitudes (60°–90°) when people are most likely to be exposed to harmful levels of solar UV and so seek shade, the erythemally-effective UV from direct UV radiation is about 83% of that from the diffuse component. As the sun falls lower in the sky, this percentage falls, but at lower solar altitudes UV exposure is less of a risk and people will be less inclined to seek shade.

If the direct component of sunlight is occluded by a barrier that has a UV transmittance of t and with a given *sky view factor* of SVF, then the protection factor (PF) provided by the shade structure can be estimated as:

$$PF = 1.83/(0.83 \times t + SVF)$$

with a corresponding transmission of 1/PF.

The sky view factor (SVF) varies between 1 for an uninterrupted sky view and 0 for a totally obstructed sky. This expression assumes the radiance of diffuse radiation is equal from all parts of the sky which, although not strictly true, is an adequate approximation for the purpose of estimating the protection from shade. Furthermore, surface albedo is not accounted for but this is very small for almost all ground covering with the exceptions of sand and snow.

As an example, we estimate the PF for a beach umbrella which has a UV transmittance of 1%, a diameter of 1.6 m and is at a height of 1 m above the ground. The SVF is calculated on a horizontal surface located vertically underneath the centre of the umbrella, mimicking the UV exposure received by someone stretched out underneath. For a simple structure such as this example, the SVF can be estimated by solid geometry and is calculated as $1 - [\pi \times 0.8^2/1^2] / 2\pi$, which equals 0.68. This results in a PF of 2.6, which corresponds to a transmission of 38%.

This value compares favourably with measurements made around the middle of the day in mid-July in Valencia (latitude 39.5°N; solar altitude 60° to 70°) with an umbrella of the same radius and height above ground, in which a mean transmission of 34% was obtained over the period of measurement [2].

Raising the umbrella so it is now 2 m above the ground reduces the PF to 1.9. Clearly, small shade structures, such as parasols, can leave large amounts of sky visible and so provide only low UV protection.

In situations where the shade geometry is less straightforward, this simple approach is inadequate and either more complex calculations are required or, for existing structures, fish-eye lens photography in combination with a UV sun chart [3] can be used to estimate the SVF.

For structures where the UV transmission is negligible, the level of protection against sunburn, expressed by the PF, when the solar UV irradiance is high (solar altitudes ≥60°) may be estimated as 1.83/SVF (see figure 6.3).

We can see that even when in the shade of direct sunlight, 20% or less of the sky needs to be visible if the protection is to be worthwhile (PF>10).

6.1.3 Factors in the design of shade structures

In the design of shade structures it is crucial to consider the sun path, which refers to the seasonal and hourly positional changes of the sun as the Earth rotates and orbits around the Sun, in relation to the orientation of the shade structure.

Sun paths at any latitude and any time of the year can be determined from astronomical geometry, and figure 6.4 shows an example for Madrid (latitude 40.4° N, longitude 3.7°W) on 21 June when solar noon occurs at a local time of 14:16 h.

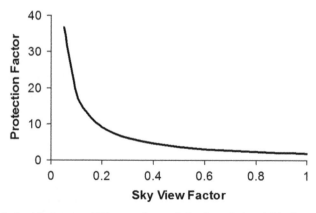

Figure 6.3. The relationship between UV protection and fraction of sky visible from a shade structure, assuming the solar disc is occluded, at solar altitudes above 60°.

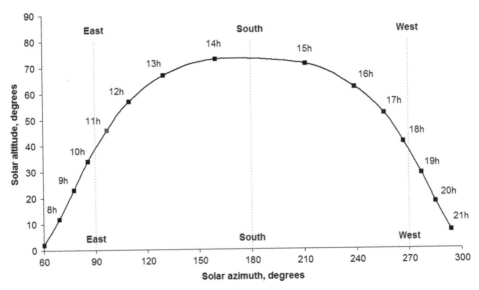

Figure 6.4. The sun path in Madrid on mid-summer day. Time is given in local (daylight saving) time, which is two hours ahead of GMT, and the solar azimuth angle is expressed relative to true north.

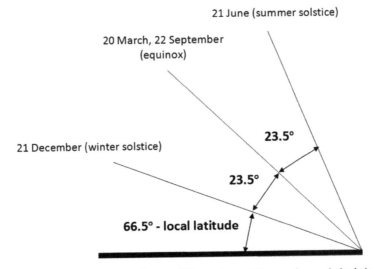

Figure 6.5. The differences in peak solar altitude at different times of the year for any latitude in the northern hemisphere.

The Earth's axis of rotation tilts about 23.5°, relative to the plane of the Earth's orbit around the Sun. As the Earth orbits the Sun, this creates a 47° peak-to-peak solar altitude angle difference between mid-winter and mid-summer (figure 6.5).

We can infer from figure 6.5 that on the winter solstice at latitudes more northerly than 66.5°N, the sun never rises above the horizon. At a latitude of 23.5°N, the sun is directly overhead (solar altitude 90°) on the summer solstice but at locations closer to the equator, peak solar altitude does not occur on this date but on two other occasions. For example, at Kuala Lumpur (latitude 3.1°N), the peak solar altitude is 70° on 21 June but reaches 90° on 27 March and 15 September.

Some architectural design software includes elements on sun paths and is able to plot building and canopy shadows for any geographical location throughout the day and year.

Shading material comes in various forms that includes solid material such as wood that offers complete protection, trellis and netting where the mesh size impacts on the protection achievable, UV-opaque but visible-light transmitting plastics, cloth where the weave density determines the level of UV transmission, and foliage or vines.

In designing shade, factors other than ambient UV levels may need to be considered, such as expected air temperature, wind speed and likelihood of rain. In some locations for some of the year, UV levels may be sufficiently high to warrant shade but temperatures are too low for comfort and this might be exacerbated by cooling sea breezes. Such locations may require 'warm shade'; the use of a shading material which transmits heat but blocks UV, such as laminated glass or polycarbonate.

A common application for shade not only against solar UV radiation, but also to provide protection against heat, wind and rain is the residential conservatory in which the living space is brought into the outdoors. Figure 6.6 shows an example where the roof material is a PVC-coated fabric, which combines high translucency with excellent UV protection.

6.1.4 Shading at schools

Children may be outside for extended periods at school either in lunch breaks or for sporting activities. Not only is sunburn a risk during the summer months but sun

Figure 6.6. Outdoor living space with a UV protective, PVC-coated fabric roof (courtesy of Fresco Shades, Auckland, NZ).

exposure in childhood is a critical period for subsequently developing melanoma. Consequently, providing shade for children whilst at school is a high priority, especially in countries of high insolation such as Australia and New Zealand.

Shade may need to be available in a number of different areas of the school, that can include the main entrance, covered walkways, verandahs and porches, communal areas where pupils may gather for lessons or lunch, and recreational areas such as sports fields and swimming pools [4].

An example of communal shade that provides a large area of UV protection and which is used for a number of purposes such as eating lunch, group activities and free play, is shown in figure 6.7.

Figure 6.8 shows a different example in which deciduous vines over a pergola create dense, cool shade that is important in countries such as Australia and New Zealand where the summers can be very hot.

6.1.5 Shade from trees

Trees provide a natural means of shade, but the diversity of genus and tree canopy dimensions make it difficult to generalise the degree of protection afforded. Protection factors for single trees are typically in the range from around 4 to 10 depending on foliage and proximity to the periphery of the shadow cast by the tree, although underneath a tree with dense foliage and low hanging branches, such as fir trees, protection factors can be 20 or higher. In thickly wooded areas where exposure to the sky is virtually eliminated, the degree of protection from solar UV may be in excess of 50-fold.

In general, the protection provided by trees increases with width of the tree canopy, low hanging branches, proximity to the trunk and density of foliage.

Figure 6.7. Communal shade at school provided by a canopy (courtesy of Christina Mackay, Victoria University of Wellington, New Zealand).

Figure 6.8. Shade at school using natural vegetation (courtesy of Christina Mackay, Victoria University of Wellington, New Zealand).

6.2 Clothing

The protection provided by clothing is a binary system; covered body sites are protected to a degree determined by the fabric's UV transmission, whereas uncovered sites remain unprotected. Clothing is an effective and reliable source of protection against solar UV radiation, provided the garment exhibits good coverage of the skin and the fabric prevents most of the incident UV radiation from reaching the skin beneath it.

6.2.1 The nature of fabric

The nature of the fabric used in a garment is at the core of its UV-protective properties.

Fabrics are manufactured from fibres, which may be *natural* e.g. cotton, linen, and wool, or *artificial* e.g. rayon, acetate, nylon, acrylic and polyester. Much clothing worn in the summer months is made entirely or partially from cotton, or a blend of polyester with cotton. Clothing that is sold specifically for its UV-protective properties, e.g. beachwear, and that makes a claim of preventing sunburn, is generally made from artificial fabrics.

Fibres, whether natural or artificial, are made into yarns, which are cylindrical structures whose length is considerably greater than their width. The yarns are then used to manufacture fabric, either as knitted fabrics composed of intermeshing loops of yarn, or woven fabrics that have two or more sets of yarn interlaced at right angles to each other.

6.2.2 Factors affecting UV protection of fabrics

A number of factors affect the protection offered by fabrics against solar UV radiation; these include yarn density, composition of fibres, colour, thickness, wetness, and laundering.

Yarn density

Whilst factors such as colour or thickness of material influence a fabric's photo-protective properties, it is the presence of interstices, or 'holes', that is the most important. The tighter the knit or weave, the smaller the interstices and the less UV that is transmitted (see figure 6.9)

Composition of the fibres

Fibres will naturally absorb some UV radiation in an analogous way to the active ingredients in sunscreens. Different fibre types absorb to different extents with polyester being the most photoprotective, followed by wool, silk and nylon, with cotton and rayon providing the least photoprotection.

Colour

Darker colours of a given material absorb more UV than lighter colours. Many dyes that are used to colour fabrics will act to increase the absorption of UV and so reduce exposure. However, colour alone is a poor guide to UV protection.

Thickness

As expected, heavier, thicker clothes attenuate more UV; for example, a sheer silk blouse provides far less UV protection than cotton denim jeans. In addition, fabrics that stretch easily, such as lycra, increase their UV transmission when stretched compared with the relaxed state

It is apparent from simple observation that ladies' stockings are transparent to light. The standard measure of yarn thickness for stockings is *denier*, defined as the weight in grams in a 9000 m length of fibre and is the industry standard in hosiery. Similar to visible light, measurements confirm UV transparency of plain knit stockings; stockings in the range 40 to 10 denier transmit between 30%–70% of incident UV, with the most popular type of stocking (15 denier) transmitting over 50% irrespective of colour. This property is likely to be a contributory factor to the higher incidence of skin cancers and greater degree of elastosis on the lower leg of

Figure 6.9. Five examples of fabrics, all with different density of yarn per unit of surface area and providing different levels of UV protection. The average UV transmissions going from left to right are approximately 30%, 10%, 3%, 2% and 1% (© Commonwealth of Australia 2017 as represented by the Australian Radiation Protection and Nuclear Safety Agency (ARPANSA)).

women compared with men, and women who require photoprotection to the legs are advised to wear trousers or at least a 40 denier stocking.

Wetness

A wet garment may, but not always, exhibit increased UV transmission. Studies have shown that, on average, the protection factor of a wet cotton T-shirt may fall to as low as one-half of the protection when dry. This effect is much less with lycra or elastane garments that are designed to be worn when swimming.

Laundering

When garments are laundered the fabric may shrink and so increase its UV protection. Cotton and rayon fabrics are more likely to shrink than nylon or polyester.

6.2.3 Determination of the photoprotection provided by fabrics

One consequence of the increasing concern about sun exposure has resulted in methods to determine the UV protection afforded by clothing fabrics and the introduction of the *ultraviolet protection factor (UPF)*, which is analogous to the *sun protection factor (SPF)* associated with sunscreens (see section 7.3).

The usual method of determining the UPF is to use an *in vitro* assay. Briefly, this method utilises a source of UV radiation and a spectroradiometer to measure the UV intensity on a wavelength-by-wavelength basis, after and before passing through the fabric sample. The spectral transmission, $T(\lambda)$, of the fabric at wavelength λ nm is simply the ratio of these two quantities, and by combining this with the CIE (1998) reference erythema action spectrum ($S(\lambda)$, black curve in figure 4.9) with a reference solar spectrum ($E(\lambda)$, table 3.2), the *UPF* is determined as:

$$UPF = \sum_{290}^{400} E(\lambda)S(\lambda)\Delta\lambda \Big/ \sum_{290}^{400} E(\lambda)S(\lambda)T(\lambda)\Delta\lambda$$

The spectral absorbance of a selection of clothing fabrics is shown in figure 6.10.

We see from this figure that most clothing fabrics, like natural shade, provide us with a spectral absorption profile that is approximately flat across the UV spectrum.

The examples shown in figure 6.10, result in UPFs that range from about 10 to over 300.

6.2.4 Summary of ultraviolet protection factors (UPFs)

A number of laboratories in several countries determine UPFs on (normally) summer-weight clothing fabrics. Probably the laboratory that has the largest database of results is the Australian Radiation Protection and Nuclear Safety Agency (ARANSA) and figure 6.11 summarises the distribution of UPFs measured on almost 29 000 samples.

From these data we find that about two-thirds of samples have a UPF of 50 or higher and more than 95% have a UPF greater than 10.

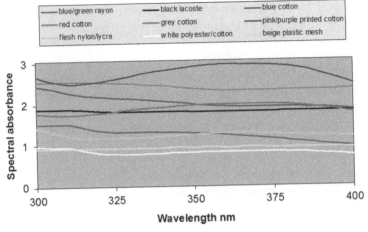

Figure 6.10. The spectral absorbance of a selection of clothing fabrics in a dry, relaxed state (data courtesy of Dr HP Gies, ARPANSA). Courtesy of Australian Radiation Protection and Nuclear Safety Agency.

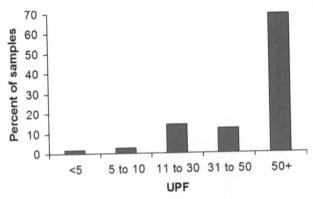

Figure 6.11. Distribution of UPFs measured on 28 849 samples (data courtesy of Dr HP Gies, ARPANSA). Courtesy of Australian Radiation Protection and Nuclear Safety Agency.

Although all of the major types of fabrics had measured UPFs ranging from less than 5 to greater than 30 000, it was observed that the median UPF for nylon/elastane fabrics was around 160, about twice the value for cotton and polyester fabrics.

6.2.5 Clothing UPF standards

In 1996, Australasia pioneered a method for evaluating the sun-protective capabilities of clothing based on the transmission of UV radiation through fabric [5]. The standard requires that UV protection claims can be made with a maximum limit of UPF 50+, even though many of the sun protective fabrics have measured UPFs well in excess of this. An upper limit was adopted to prevent escalating claims from clothing manufacturers. The Standard includes a UPF classification system and garments that satisfy a given rating are entitled to display a swing tag to this effect, as illustrated in figure 6.12.

Figure 6.12. A garment in an Australian store displaying a swing tag confirming that it provides UPF 50+ protection.

The original Australian/New Zealand Standard did not consider the body surface area coverage of garments although this requirement was incorporated into some subsequent standards, such as those of the European Union [6]. The Australian/New Zealand Standard has recently been revised and like the European standard, it now requires upper-body garments to cover three-quarters of the upper arm, and the torso from the base of the neck to the hip-line, while lower-body garments must cover from the hip-line to at least mid-way between the crotch and the knees.

6.3 Hats

Hats not only protect the scalp, which is a common site for actinic lesions in balding men, but can provide shade to parts of the face with the extent of shade depending on the hat design. The material from which hats are made will almost always exhibit a high UPF and so protection to the scalp, and often the forehead, generally exceeds 50-fold.

Most styles of hat are mainly protective against the direct sun, especially at high solar altitudes ($>60°$) when the erythemal radiation is at its most intense and the area of the face in shadow is at its greatest. However, at low solar altitudes, or in cloudy conditions when diffuse UV radiation constitutes a major fraction of terrestrial UV, hats become less effective and to ensure good facial photoprotection, may need to be worn in conjunction with sunscreen.

The photograph of the three styles of hats shown in figure 6.13 was taken when the solar altitude was $40°$ and we see that it is only parts of the face above the mid-nose that are in shade.

A wide-brimmed hat, such as the Panama hat shown in the left-hand panel of figure 6.13, offers the best type of facial protection with typically five-fold protection

Figure 6.13. Three types of men's hats worn in the summer: Panama hat (left); bucket hat (centre); baseball cap (right).

to the nose and ears, 2–3-fold to the cheeks, falling to 1–2-fold to the chin and neck, but only when these areas are in shade of direct sunlight. Bucket hats (figure 6.13 centre) perform almost as well providing the hat has a brim width of 5 cm or more. Baseball caps (figure 6.13 right) provide good protection to the nose but leave the ears, cheeks and neck unprotected. Legionnaire style hats, with a flap of fabric covering the neck and ears, are particularly effective at protecting these sites.

In summertime, hats are worn most often by children, are least popular with adolescents, with adults somewhere in between.

6.3.1 How wearing a hat can be assessed in terms of health benefit

Suppose we wish to look at the effect of wearing a hat on the incidence of solar keratoses, a pre-cursor of skin cancer, in bald men who work outdoors. One group of men, randomly selected from the study population, always wear a hat whenever they are at work—we refer to these men as the *treatment* group. The other group, referred to as the *control* group, never wear a hat outdoors at work. Simple random allocation such as this can be refined by using *stratification*, which is controlling for factors such as age and skin type that are known to influence the likelihood of developing an actinic keratosis.

At the end of the study period the number of men in each group who have developed at least one solar keratosis is counted. From these data we can construct a table that describes the frequency of the two possible outcomes for each of the two groups.

	Number of men who developed solar keratosis	Number of men who did not develop solar keratosis
Treatment group	a	b
Control group	c	d

The probability of an event in the treatment group is $a/(a+b)$, that is the number of men who developed a solar keratosis out of the total number of men wearing a hat. Likewise, the probability of an event in the control group is $c/(c+d)$.

The ratio of these two probabilities is the relative risk and is expressed as:

$$\frac{a/(a + b)}{c/(c + d)}$$

If wearing a hat did reduce the likelihood of developing a solar keratosis, the relative risk should be smaller than one, but if $RR = 1$, wearing a hat made no difference at all. If RR is above 1, then wearing a hat actually increased the risk of developing a solar keratosis.

When expressing the risks from behaviour and intervention, the terms *relative risk* and *odds ratios* are often confused and the terms sometimes used interchangeably. The basic difference is that the odds ratio is a ratio of two odds, whereas the relative risk is a ratio of two probabilities.

In the example here, the odds of an event in the treatment group is the number of hat-wearing men who developed a solar keratosis divided by the number of hat-wearing men who didn't develop the lesion; that is, a/b. Similarly, the odds for the control group is c/d.

The odds ratio is then expressed as:

$$\frac{a/b}{c/d} = ad/bc$$

When the number of events (solar keratoses) in each group is small compared with the number of subjects in the respective groups (treatment and control), we see that the relative risk and odds ratio are approximately equal.

6.4 The changing fashion for summer clothing

At the turn of the 20th century summer leisure wear was exemplified by the type of outfits shown in figure 6.14.

This style of long dress accompanied by large hats was popular for wear at summer garden parties and fêtes. The gentleman's suit, consisting of matching coat, trousers and waistcoat, was accepted attire for summer sports and seaside wear. The outfit was often completed with a straw boater. Clearly both the male and female fashions prevalent at the time provided excellent sun protection.

In terms of dramatic turning points, the 1920s were a key time in the development of summer wear as skirts became shorter, dresses were lightweight with often plunging necklines and the popularity of suntanning took hold.

In the early years of the 20th century, swimsuits were one-piece with only the arms and lower legs exposed, but by the 1920s beachwear became quite revealing, and apart from the fact that swimsuits had a slightly longer cut in the leg and were made from thick knitted wools and cottons, they were not too dissimilar from swimsuits today.

Men's bathing suits of the time were less revealing than modern swimwear and often consisted of a singlet top and shorts.

The rise of the bikini in the 1950s was another important milestone in beachwear as suddenly it became acceptable to expose the midriff. The American designer Rudi

Figure 6.14. A sketch of outdoor leisure wear worn by fashionable men and women around the turn of the 20th century.

Gernreich's topless swimsuit of 1964 was very *avant-garde* and not something that ever really took off but within a decade, topless sunbathing was a regular activity on European beaches.

By the 1980s, with the arrival of the thong bikini, a point was reached where anything was acceptable in terms of tiny bikinis (and speedos for men), as well as a general acceptance of the naked body in public spaces.

Over the past 20–30 years, whilst there have been stylistic variations, no great taboos are waiting to be broken. An interesting development in women's swimwear has been the burkini, which covers the whole body except the face, the hands and the feet. The design is intended to accord with Islamic traditions of modest dress (figure 6.15).

Towards the last decade of the 20th century people were becoming more aware of the harmful effects of excessive sun exposure and specially-designed UV-protective swimwear became available, especially in sunny countries like Australia. Figure 6.16 shows examples of sun-protective swimwear, which cover the whole trunk, as well as the upper arms and legs. The style is very much a throwback to earlier times although the fabrics used are modern, synthetic materials such as lycra rather than wool, which was the material of choice for swimwear a century ago.

6.5 Sun protection accessories

Examples of items that provide sun protection for specific purposes are illustrated in figure 6.17.

Figure 6.15. Example of burkini swimwear (courtesy of EastEssence.com, California).

Figure 6.16. Modern sun-protective swimwear (courtesy of Angeli Jackson, SunSibility, Epsom, UK).

The palmless gloves are ideal for gardening or driving where the backs of the hands are protected but still allowing tactile sensation.

The nose is a frequent site for skin cancers and protecting this vulnerable area by clipping the guard onto the bridge of sunglasses may appeal to some people. And for the cyclist who wants extra protection to the neck, ears and scalp (important for bald men), a slip-on fabric with high UPF attaches to the cycle helmet creating a virtual legionnaire cap.

6.6 Estimation of body surface area covered by clothing

The area of the body surface covered by clothing is an important factor in assessing a potential risk from sun exposure. A number of approaches have been proposed ranging from the 'rule-of-nines', a simple formula first developed to assess the extent of a patient's burns, through to dividing the body surface into 142 regions based on anthropometric standard points [7].

An approach somewhere between these two extremes is given in table 6.1 in which the regions have been specifically selected to reflect typical areas covered by clothing.

As an example, table 6.2 illustrates the percentage of body surface area covered by clothing in a man wearing a short-sleeved shirt, trousers and shoes, and a woman wearing a sleeveless, knee-length dress and sandals. In the case of sandals an estimate is made of the area of foot that is covered.

Although not shown specifically in table 6.1, a man wearing only a pair of swimming trunks or a woman wearing a skimpy bikini would both have about 93% of their body surface exposed.

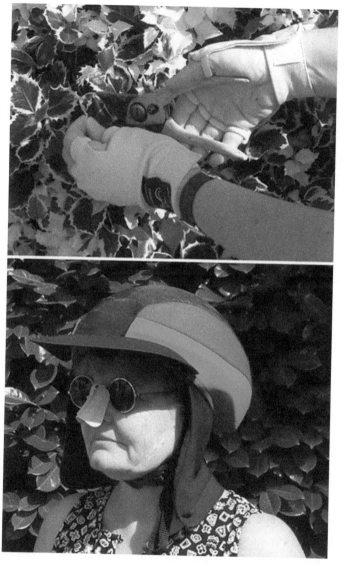

Figure 6.17. Palmless gloves (top), nose guard and cycle helmet protector with neck flap (bottom).

6.7 Optical filters

Most people spend the majority of their time indoors, either in buildings or in vehicles. That our interior space is illuminated is a consequence of the high transmission of visible radiation through glass or plastics—our interface with the outside world. In this section we examine the transmissivity of these materials for solar UV radiation.

Table 6.1. Regional body surface area (BSA) in adults.

Region	% BSA
Head, neck and ears	9.0
Upper trunk—bottom of neck to waist	15.3
Lower trunk—waist to bottom of pelvis	20.6
Upper arms—shoulders to elbows	8.8
Lower arms—elbows to wrists	6.1
Hands	4.8
Upper legs—hips to knees	15.3
Lower legs—knees to ankles	13.6
Feet	6.5
Total	100

Table 6.2. Percentage of body surface covered by clothing in a man wearing a short-sleeved shirt, trousers and shoes, and a woman wearing a sleeveless, knee-length dress and sandals.

Region	Man	Woman
Head, neck and ears	0	0
Upper trunk—bottom of neck to waist	15.3	15.3
Lower trunk—waist to hips	20.6	20.6
Upper arms—shoulder to elbow	8.8	0
Lower arms—elbow to wrist	0	0
Hands	0	0
Upper legs—hips to knees	15.3	15.3
Lower legs—knees to ankles	13.6	0
Feet	6.5	4.5
Total	80	56

6.7.1 Nature of glass and plastics

Glass is made by melting together silica in the form of sand with other minerals, such as soda ash and limestone, at temperatures of around 1700 °C. Other materials can be added to produce different colours or properties. The types of glass found in residential and commercial buildings include clear, tinted, reflective, low-emissivity, tempered, insulating and laminated. The thickness of glass used for architectural purposes is generally in the range 3 to 6 mm.

Polycarbonate is a naturally transparent amorphous thermoplastic, with the ability to transmit light almost to the same extent as that of glass. It has high strength, toughness, heat resistance, and excellent dimensional and colour stability. Solar UV radiation will degrade polycarbonate and when the material is intended for external use, UV stabilisers are normally added to reduce the rate of photodegradation.

An economic alternative to polycarbonate when extreme strength is not necessary is poly(methyl methacrylate) (PMMA), also known as acrylic or acrylic glass as well

Figure 6.18. A canopy with a visibly-clear, UV-blocking roof constructed of 16 mm polycarbonate (courtesy of NBB School Shelters, Poole, UK).

as by trade names that include Plexiglas®, Acrylite®, Lucite®, and Perspex®. PMMA is a transparent thermoplastic often used in sheet form as a lightweight or shatter-resistant alternative to glass.

6.7.2 Spectral transmission of glass and plastics

Materials that are visibly clear will absorb UV radiation to varying extents. Clear window glass transmits UV radiation down to about 310 nm, whereas polycarbonate does not transmit below 370 nm. For this reason, shade structures that need to allow through visible light but provide high attenuation against erythemal UV radiation will often have roofs constructed from polycarbonate (figure 6.18)

PMMA can be formulated to absorb virtually all UV radiation or alternatively, to provide high transmittance from wavelengths of about 290 nm upwards, which has made it suitable as a substrate for the *in vitro* assay of sunscreens (see section 7.4).

6.7.3 Glass windows in automobiles

Windscreens on cars and other vehicles are made from a laminate of glass and plastic so that, if involved in a collision, the glass fragments will adhere to the plastic interlayer rather than break free. Side and rear windows, on the other hand, are made from toughened (non-laminated) glass, apart from some high-end luxury cars. Both types of glass are usually tinted to improve comfort by reducing the transmission of visible and infrared radiation.

Figure 6.19 shows representative transmission spectra for various glass windows found in cars. We see that laminated glass is virtually opaque at wavelengths below 370 nm, whereas non-laminated glass will transmit down to about 310 nm depending on the tint.

6.7.4 Clinical impact of UV transmission through car windows

In people with normal responses to sunlight, travelling in vehicles with the windows closed is unlikely to present a problem since almost all erythemal UV radiation (mainly UVB) will be blocked by the windscreen and side windows.

However, patients with photosensitivity disorders, such as polymorphic light eruption (PLE), who spend long periods in vehicles often report provocation of their rash. This can be explained from figure 6.19, which shows that the UVA waveband, which precipitates the rash, is readily transmitted through side windows, but not the laminated windscreen.

In addition to lamination and tinting, other factors influencing UVA exposure within a vehicle include the position of the individual, direction of travel with respect

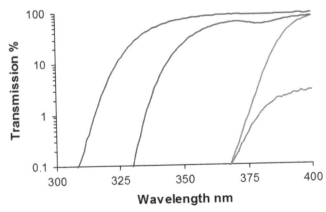

Figure 6.19. The spectral transmission of representative samples of car window glass:grey laminate (grey); clear laminate (red); non-laminated light green (green); non-laminated clear (blue) (data courtesy of Dr Jim Lloyd).

to the sun, time of day and cloud cover. Studies have shown appreciable differences in the exposure at various anatomical sites within a car. This is especially true of the forearms, a common site for developing PLE, with the forearm furthest away from the window receiving 1% or less of ambient UV radiation outside the car. If the arm closest to the (closed) window is resting at a level below the bottom of the side window and is in shadow from direct sunlight, the maximum exposure is typically around 5%–10% of ambient with the side window facing towards the sun.

The worst case occurs when the arm nearest the window is elevated to the extent that it is exposed to direct sunlight. Here, the exposure through a closed window of non-laminated clear glass will be around two-thirds of ambient, whereas with the window open the exposure may be higher than the ambient on a horizontal plane outside the car since the arm might be perpendicular to the sun's rays.

There are anecdotal reports that solar keratoses appearing on the arms are more frequent on the right side of people in Australia and the UK, and on the left side of people in the US. This is said to be due to the greater amount of sunlight received on the driver's side when driving motor vehicles with the window open, since the arm next to the open window will be exposed to considerably higher UV levels than the contralateral arm.

6.8 Sunglasses

Sunglasses are commonly worn not just to reduce the glare from bright sunlight but also to protect the eyes against UV damage (section 4.6).

Sunglass lenses are made of either glass or plastics that include acrylic, polycarbonate, CR-39 or polyurethane. Whilst glass lenses have the best optical clarity and scratch resistance, they are heavier than plastic lenses and can shatter or break on impact. For this reason, plastic lenses are generally preferred since they are lighter and shatter-resistant, but they are more prone to scratching. CR-39 is the most common plastic lens used for sunglasses due to its low weight, high scratch resistance, and low transmittance of UV radiation. Sunglasses that transmit 1% or less of UV radiation at wavelengths up to 400 nm are often labelled as *UV400*.

As well as choice of lens material, the style of sunglass is also important for optimal eye protection. Sunglasses that wrap around at the side, or have windows in the side arms, are preferred as they offer protection against scattered or reflected UV radiation that can reach the eye. UV entering from the side can be focused to the medial side of the eye close to the nose, a phenomenon referred as the *Coroneo effect*. This site is where pterygium preferentially occurs.

6.8.1 Sunglass standards

The international standard for sunglasses is ISO 12312, which was published in 2013. Part 1 specifies the physical and optical characteristics of glasses, including a range of UV protection levels, while Part 2 specifies the test methods used to validate conformance with Part 1. Other sunglass standards include those from Australasia (AS/NZS 1067:2003 Sunglasses and fashion spectacles), the British standard (BS EN

Figure 6.20. Children wearing brimmed hats in the shade of a tree (courtesy of Christina Mackay, Victoria University of Wellington, New Zealand).

ISO 12312-1:2013+A1:2015. Eye and face protection. Sunglasses and related eye-wear. Sunglasses for general use), and the US standard (ANSI Z80.3-2001).

6.9 Combining protection modalities

Some modalities may only provide very modest photoprotection. For example, a brimmed hat may only deliver three-fold protection to the cheeks, and a tree with open foliage could offer just five-fold protection. But wearing a brimmed hat, whilst in the shade of a tree, as illustrated in figure 6.20, would provide worthwhile protection, since the overall protection factor is approximately the product of the individual protection factors, which in this example would be 3 multiplied by 5, or 15-fold.

If the children in figure 6.20 were also wearing sunscreen with an effective SPF of 5, taking into account average application thickness and uniformity of spreading, then their cheeks would be roughly protected to 15 multiplied by 5, or 75-fold relative to exposure in full sun with no hat or sunscreen.

References

[1] Parisi A V and Kimlin M G 1999 Comparison of the spectral biologically effective solar ultraviolet in adjacent tree shade and sun *Phys. Med. Biol.* **44** 2071–80

[2] Utrillas M P, Martínez-Lozano J A and Nuñez M 2010 Ultraviolet radiation protection by a beach umbrella *Photochem. Photobiol.* **86** 449–56

[3] Kawanishi T 2013 Sky erythema ultraviolet radiance and UV shade charts. Radiation processes in the atmosphere and ocean (IRS2012) *Proc. of the International Radiation Symposium (IRC/IAMAS), AIP Conf. Proc.* **1531** 876–8

[4] Mackay C 2003 Sunshade design in New Zealand primary schools 2003 *The 20th Conf. on Passive and Low Energy Architecture (Santiago, Chile, 9–12 November 2003)*

[5] Standards Australia/Standards New Zealand AS/NZS 4399 1996 Sun protective clothing—evaluation and classification (Sydney) (Wellington: Standards Australia) (Standards New Zealand)

[6] European Standard EN 13758-2 2003 Textiles: solar UV protective properties—Part 2: *Classification and marking of apparel* (Brussels: CEN)

[7] Lee J Y and Choi J-W 2009 Estimation of regional body surface area covered by clothing *J. Human-Environ. Syst.* **12** 35–45

Sun Protection
A risk management approach
Brian Diffey

Chapter 7

Sunscreens

In the previous chapter, physical barriers to protecting the skin from solar UV exposure were covered. Here we examine the most well-known and widely-used chemical barrier—topical sunscreens.

7.1 The nature of topical sunscreens

A topical sunscreen is a substance applied to the surface of the skin in order to reduce the intensity of solar UV radiation entering the skin and so mitigate damage to vulnerable cells in the epidermis and dermis. Sunscreens can take many forms including creams, milks, lotions, gels, foams, oils, ointments and sprays. The active ingredients of a sunscreen are normally between one and six (and occasionally more) chemicals that can either absorb UV radiation and then dissipate the energy in the form of heat or phosphorescence, or scatter the incoming UV radiation away from the skin.

7.1.1 Active ingredients

The heart of any sunscreen product is the active ingredient(s), commonly referred to as the *UV absorber* or *UV filter*. UV filters may be either inorganic or organic chemicals. There are currently about 55 UV filters that are approved by various agencies for use in sunscreen products: 16 UV filters are approved for sunscreen products in the United States, 20 in Canada, 28 in the European Union, 28 in Asia, and 33 approved in Latin America. However, only 11 of them have been approved globally and these are listed in table 7.1 together with their absorption maximum (λ_{max}) to give a guide as to whether they may be considered UVB or UVA filters.

Inorganic UV filters
There are only two inorganic UV filters permitted in sunscreen, which are titanium dioxide and zinc oxide. Both these chemicals are available in micronised and nanosized forms (sometimes called nano-particles), which give them good sun

Table 7.1. UV filters that are universally approved for use in topical sunscreens.

Colipa No.	INCI	USAN	λ_{max} (nm)
S1	PABA	PABA	290
S8	Ethylhexyl dimethyl PABA	Padimate-O	311
S12	Homosalate	Homosalate	306
S20	Ethylhexyl salicylate	Octisalate	305
S28	Ethylhexyl methoxycinnamate	Octinoxate	311
S32	Octocrylene	Octocrylene	303
S38	Benzophenone-3	Oxybenzone	324
S45	Phenyl benzimidazole sulphonic acid	Ensulizole	302
S66	Butyl methoxydibenzoylmethane	Avobenzone	357
S75	Titanium dioxide (TiO_2)	Titanium dioxide	>290
S76	Zinc oxide (ZnO)	Zinc oxide	~370

Abbreviations: Colipa—The European Cosmetics, Toiletry and Perfumery Trade Association (now renamed Cosmetics Europe); INCI—International Nomenclature of Cosmetic Ingredients; USAN—United States Adopted Name

protection properties without imparting the traditional opaqueness that is aesthetically unappealing in cosmetic formulations. Inorganic UV filters are generally combined with organic UV filters in sunscreens to achieve high SPFs.

Inorganic chemical filters are sometimes referred to as *physical blockers*, but this term should be avoided, as should the term *sunblock*, as clearly no topical product applied to the skin can provide complete attenuation of incoming UV radiation.

Organic UV filters

Organic chemical absorbers are often classified into either UVA or UVB filters depending on the type of radiation they largely absorb. Examples of UVA absorbers include Benzophenone-3 and Butyl methoxydibenzoylmethane, whereas UVB absorbers include Ethylhexyl methoxycinnamate, Octocrylene and PABA (see table 7.1).

Recent developments in UV filters have high molecular weights (over 500 Daltons) to diminish their penetration into the skin. These molecules possess multiple chromophores that yield high extinction coefficients and also broad-spectrum protection. Examples include Bis-ethylhexyloxyphenol methoxyphenyltriazine S81 (Tinosorb S, BASF, Germany) and Drometrizole trisiloxane S73 DTS (Mexoryl XL, L'Oreal, France). These more effective UV filters are, unfortunately, not yet (as of 2017) approved in all countries.

Photostability

The exposure of UV-absorbing molecules to solar UV radiation may lead to photochemical reactions that can compromise both physical attributes, such as colour and appearance, and chemical properties leading to undesirable reactions and by-products. A molecule that reacts in this way is termed *photounstable* and a widely-used example is avobenzone. Approaches to improve the photostability of UV filters include the use of glass beads and microspheres, as well as UV-absorbing quenching molecules that include octocrylene and 4-methylbenzylidene camphor.

Other ingredients

In addition to the active UV filters, there will be several other chemicals called *excipients*, such as emulsifiers and surfactants, that are used to formulate the product and give the product its particular cosmetic feel, as well as contributing to its performance in terms of water resistance. Also, some sunscreens may contain antioxidants that can play some role in reducing UV damage to the skin, as well as preservatives that inactivate detrimental bacteria, yeast and/or moulds.

7.2 Sunscreen use

The first use of sunscreens was reported in 1928 in the USA and they were originally intended not so much to protect the skin from harm but to encourage tanning by reducing the risk of burning, as the advertisement illustrated in figure 7.1 shows. Since those early days, sunscreens have become increasingly popular, principally in North America, Europe and Australia. Sunscreen use has also become common among non-white populations in Asia and Latin America.

Sunscreens are widely available for general public use as a consumer product and are mostly sold in supermarkets and in pharmacies as over-the-counter products. They are also available directly from physicians (e.g. in the USA), from hospitals (e.g. in Italy) and from cancer control organisations and cancer charities (e.g. in Australia). In Australia, sunscreens are available in the workplace as part of occupational health and safety programmes, and they are widely available in schools since their use by children is actively promoted. In contrast, schools in the United States, partly because of fear of litigation, rarely promote sunscreens as these products are classified as drugs.

Whilst the primary purpose of sunscreen is to reduce the amount of solar UV radiation reaching the viable epidermis, the underlying reasons that people use them include reducing the risk of skin cancer, preventing sunburn, promoting tanning by avoiding burns that blister, protecting the skin from photoaging, enabling outdoor activities or simply to comply with the expectations of others. Except for people who may use sunscreen routinely as a component of daily skincare, sunscreen use (like other sun-related behaviour) is a contingent rather than a habitual behaviour. That is to say, it is contingent upon certain situations, such as a hot, sunny day or the desire to spend time outdoors.

7.2.1 Why do people use (and not use) sunscreens?

In surveys examining people's beliefs about reducing the risk of skin cancer, the measure regarded most widely as being very important by almost everyone is the use of sunscreen. The other measures asked about—avoiding the midday sun, staying in the shade, and wearing clothing and a wide-brimmed hat—are generally considered less important.

Figure 7.1. A 1930s advertisement for a sunscreen (Wellcome Library, London).

By far the most common reason that people give for using sunscreen is to protect against sunburn. Other reasons given include:
- Know dangers of sun exposure.
- Perceive themselves at high risk of skin cancer.
- Know people who had skin cancer.
- Protect against ageing and wrinkling.
- Extend time in the Sun.
- Previously had skin cancer.

We should not forget that many people do not use sunscreens regularly, or at all, and the reasons that people give for not choosing to apply sunscreens to their skin include:

- Have skin that does not burn easily.
- Already have a 'protective' tan.
- Takes too much time to apply.
- Not outdoors enough to warrant use.
- Nuisance and greasy to apply.
- Feels hot and sweaty.
- Expensive.
- Retards desired tan.
- Use other sun protective measures.
- Forget.

When their application is commensurate with exposure, sunscreens undoubtedly protect against sunburn, although it is a truism that many people who apply sunscreen in the belief that they are protected will develop subsequent sunburn, as is discussed in more detail in section 7.7.

7.2.2 Demographics of sunscreen use

There is great variation in sunscreen use within individuals according to their natural sun susceptibility, socioeconomic status, propensity for sun exposure, holiday habits, perception of hazards associated with sunscreen use, and strength of the Sun.

Although the prevalence of sunscreen use has been reported in countless studies, no standard metric has been used to quantify use and in a number of studies in which data on sunscreen use were collected, they were not reported separately from other indices of sun protection, into which prevalence of sunscreen use was merged. There is thus considerable variation in the way in which data are reported and because of this heterogeneity, summary metrics should be treated with caution.

It is difficult to assess actual sunscreen application from direct observation and only a few studies have tried to assess prospectively real-life sunscreen use patterns. Most data on sunscreen use proceed from surveys directly asking people whether they used a sunscreen when in the Sun.

In some studies, a point prevalence of use is reported: for instance, people interviewed on the beach may be asked 'Are you using a sunscreen now?' and for people in a telephone survey a question might be 'Were you using a sunscreen between 11:00 am and 3:00 pm yesterday?'

A question about typical use is often asked, such as 'In summer, how often do you use a sunscreen when out of doors?' It must be borne in mind, however, that sunscreen use assessed through questionnaires is subject to bias and over-reporting.

In all the studies in which sex differences were reported, use was greater among women than men. Another feature is the consistency of reported use by age group in that children's use of sunscreen is higher than that of adolescents or adults. Not surprisingly, sunscreens are most often used during intentional exposure to the Sun, such as at the beach, and the prevalence of usual use in these studies is higher than at other times e.g. when at work or weekend exposure.

Whilst children are the largest users, sunscreen application is best avoided in infants less than six months of age as babies' skin is thinner than that of adults, and it can absorb the UV-active ingredients in sunscreen more easily. Also, infants have a high surface-area to body-weight ratio compared to older children and adults. Both these factors mean that a baby's exposure to the chemicals in sunscreens is greater, increasing the risk of an unwanted reaction.

The best approach is to keep infants under six months out of direct sun and in the shade as much as possible. This is especially important around the mid-part of the day when UV radiation is most intense. It is preferable to make use of clothing and shade but if there's no way to keep a young infant out of the Sun, applying high SPF sunscreen to small areas such as the cheeks and back of the hands is acceptable.

People who work outdoors are chronically exposed to solar UV radiation throughout the year. The advice from public health agencies to employers and workers is very much to focus on clothing and shade and to use sunscreen as a support, and not a primary, measure. Lifeguards are the group of workers perhaps most commonly associated with excessive sun exposure and one survey in the US found that over 60% of this group reported using sunscreen. On the other hand, farmers and other rural outdoor workers are also a high-risk group for skin cancer but here the use of sunscreen is generally much lower.

7.2.3 Trends in sunscreen use

As a crude indicator of trends in sunscreen use we can exploit the fact that millions of people worldwide use the Internet daily as a source of health information. Using Google Trends, which is an online search tool that allows users to see how often specific keywords, subjects and phrases have been queried over a specific period of time, data on a search query for the term *sunscreen* between January 2004 and December 2016 were compiled for Australia and the UK. The trends are shown in figure 7.2.

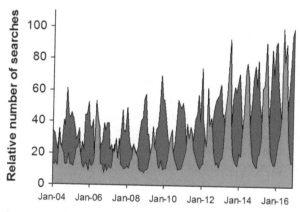

Figure 7.2. Trends in internet searches for the term sunscreen over the period January 2004 to December 2016 in the UK (red) and Australia (blue).

The ordinate represents search interest relative to the highest point on the chart for a given region and time period, with a value of 100 reflecting the peak popularity for the term.

We see that there is a strong cyclic pattern with, not surprisingly, much more interest in the summer months than the winter (note that Australian summers are six months out of synchrony with UK summers). The cyclic pattern is more pronounced for the UK than Australia, probably because of the greater winter to summer variation in ambient solar UV. Secondly, interest from UK users has increased appreciably since 2012, whereas the trend in Australia is much more steady, possibly reflecting the longer time period over which public health campaigns about the hazards of over-exposure to sunlight have been extant (see section 5.5.2).

7.2.4 Annual sunscreen use *per capita*

There are large differences in sunscreen use between continents and indeed from country to country, due to geographic location, skin colour, culture, propensity towards sun exposure, and economic status. The average annual consumptions for North America, Australasia and Europe, regions with largely white-skinned populations, are 101 ml, 60 ml, and 52 ml, respectively [1]. This equates to an average *per capita* use of 66 ml, when these quantities are weighted by the respective regional populations.

How do these annual usages compare with what we might predict? According to National Health Interview Survey data from the US, approximately one third of adults usually or always use sunscreen when outdoors in the Sun for one hour or more. As a crude estimate we assume that sunscreen is used by 30% of the population and that it is applied twice per day at an average thickness of 1 mg cm^{-2} (section 7.7.1), which is equivalent to 0.001 ml cm^{-2}.

Secondly, we make the assumption that sunscreen is applied daily during a 2 week summer holiday and on one day out of two (allowing for occasional inclement weather conditions and indoor distractions) during weekends over the 3 month summer period June to August. This results in 20 days of applying sunscreen.

Assume that the 2 week summer holiday is a mixture of beach exposure and sightseeing. During beach exposure the face, trunk, arms and legs are exposed; these areas total 90% of the body surface area (BSA). But not everyone who decides to use sunscreen will apply it to all these exposed areas and we will assume that, on average, sunscreen is applied during beach exposure to about two-thirds of the body surface.

In sightseeing, as for adventitious exposure, generally only the face, neck, forearms and hands are exposed, compromising around one-fifth of the BSA. By weighting the BSA exposed with the number of days of exposure, we estimate that, on average, sunscreen is applied to one-third of the BSA, or about 0.5 m^2, since the average BSA for all ages (children and adults) weighted by population fractions is about 1.5 m^2.

To summarise, we have:
- Prevalence of use: 30%.
- Average quantity per application: 0.001 ml cm^{-2}.

- Number of applications per day: 2.
- Number of days per year that sunscreen is used: 20.
- Area covered by sunscreen per application: 0.5 m^2 (i.e. 5000 cm^2).

This results in an annual *per capita* consumption of:

$$0.3 \times 0.001 \times 2 \times 20 \times 5000 = 60 \text{ ml}$$

This value is similar to the average *per capita* use in populations with predominantly white skin (66 ml), although it must be stressed that the assumptions is this simple analysis are very fuzzy. Nevertheless, it serves to illustrate that there is considerable potential for increasing sunscreen consumption by increasing prevalence of use, application amount and the periods throughout the year when sunscreen is applied.

7.3 The sun protection factor

The concept of the sun protection factor (SPF) was popularised by the Austrian scientist Franz Greiter in the 1970s and subsequently adopted by many regulatory authorities and the cosmetic and pharmaceutical industries. It is popularly interpreted as how much longer skin covered with sunscreen takes to burn compared with unprotected skin. So if you burn after 10 min in the Sun, then using a sunscreen labelled with, say, SPF15, is taken to mean that you can safely remain in the Sun for $10 \times 15 = 150$ min, or 2½ h, before burning.

This definition focuses on extending time in the Sun but a better way of thinking about the SPF is that if you spend a certain time in the Sun, then wearing a sunscreen with a given SPF reduces the UV exposure to the skin to 1/SPF of that you would have received by spending the same time in the Sun but with no sunscreen applied—so applying an SPF15 sunscreen results in a UV exposure to the skin of 1/15th of that you would have received if you had not applied any sunscreen. However, this statement is *only* true if the sunscreen is providing protection equivalent to the labelled SPF but as we shall see in sections 7.7–7.9, this rarely happens and most people who apply sunscreen are protected to a much lesser extent than they realise.

There is commonly-held, but fallacious, belief that applying a SPF30 sunscreen confers little additional benefit to using a SFP15 sunscreen. The basis of the argument is that SPF15 means that $1 - 1/15 = 93.3\%$ of solar UV is absorbed by the sunscreen and that SPF30 means that $1 - 1/30 = 96.6\%$ is absorbed—apparently a difference so small that it is of little benefit. But what is important to skin health is not what is absorbed by the sunscreen but how much UV reaches the skin. So in the case of an SPF15 product the skin exposure relative to unprotected skin for a given time in the Sun is $1/15 = 6.7\%$, whereas for an SPF30 product the relative exposure is $1/30 = 3.3\%$. In other words, twice as much UV reaches the skin when an SPF15 product is applied than when the same quantity of SPF30 is applied.

Thirty years ago most commercially available sunscreen products had SPFs less than 10 but by 2000 most manufacturers produced products with factors of 15 to 30, and it is not uncommon to find products today claiming an SPF of 50 or higher.

Whilst the focus in sunscreen development has been on increasing SPF, there is now a consensus in many, but not all, countries to cap at SPF 50+ .

7.3.1 History of the SPF method

Historically, the first known studies establishing the basis for the SPF, or index of protection, began more than half a century ago. These and other studies led to the first standard method for SPF determination and labelling, which was issued by the Food and Drug Administration (FDA) in the United States in 1978 [2], followed in 1984 by the DIN67501 norm in Germany [3], which was applied mainly in Europe.

These two standards differed principally in respect of the type of UV source used (xenon arc lamp or mercury arc lamp) and the surface density of product application on skin (2.0 or 1.5 mg cm^{-2}), which led to discrepancies in measured protection factors. Following agreed rationalisation, all standards issued subsequently stipulated the use of an optically filtered xenon lamp as the UV source and an application density of 2.0 mg cm^{-2}.

Standards similar to the FDA were issued by the Standards Association of Australia in 1986, which included both SPF and water resistance testing, and by the Japan Cosmetic Industry Association in 1991. Since their introduction, both the Australian and Japanese standards have undergone revision. The New Zealand Standards joined the Australian Standards for their joint new version in 1993. and their latest version was published in 2012 [4].

The European Cosmetic, Toiletry and Perfumery Association (Cosmetic Europe, formerly COLIPA), in its 1994 SPF test method, introduced new techniques to characterise and specify the emission spectrum of the UV source and to assign skin types based on skin colour rather than the subjective Fitzpatrick rating (see table 4.1). At the same time, two high SPF standard products were proposed to take into account the increase in SPF values that were appearing commercially. COLIPA, together with the standards' authorities in Japan and South Africa, began discussion on the harmonisation of the SPF measurement method in 2000 and reached a joint agreement of the International SPF Test Method in October 2002 and updated in 2006 by the European, Japanese, and South African industries [5].

The International Standards Organisation published its method for the *in vivo* determination of the SPF of sunscreen products in 2010 [6]. And in 2011, the FDA published its most recent document on the labelling and testing of sunscreens [7].

7.3.2 Measurement of the SPF

The SPF determined *in vivo* is now a universal indicator of the efficacy of sunscreen products against sunburn, and as reviewed above, detailed protocols are available.

Although outdoor testing under actual use conditions might be considered the most realistic method for SPF assay, it is both impractical and unreliable and for this reason laboratory radiation sources are used, the xenon arc solar simulator being the universal lamp of choice.

Whilst it is possible to configure the spectral output of a xenon arc lamp to provide a close match to the solar spectrum over much of the ultraviolet, visible and infrared wavebands (figure 3.15), the output power of the lamp would need to be limited so as not to exceed the small area heat pain threshold, which would necessitate exposure times of several minutes in order to elicit delayed erythema.

To overcome this limitation, solar simulators designed specifically for SPF determination incorporate a UV-transmitting, visible-light absorbing filter in the output beam in order to remove the visible light component and so allow much higher irradiances in the UV waveband compared with natural sunlight. Since wavelengths in the UVB and short-wave UVA (UVAII) region are the most effective at inducing erythema, the design goal is to match the relative shape of the spectral output of a UV solar simulator as closely as possible to sunlight in this region.

The spectral output of an Oriel® Sol-UV Series UV Solar Simulator is compared with summer sunlight in figure 7.3, where we see that, unlike sunlight, the spectral output of the UV solar simulator falls rapidly at wavelengths beyond 370 nm and emits negligible radiation beyond 400 nm.

The spectral irradiance of the UV solar simulator (UV-SS) shown in this figure is much greater in the UV region than is sunlight, as reflected by the scaling on the respective ordinates, resulting in a typical erythemal irradiance at the skin surface of about 90 SED h^{-1}, compared with mid-latitude summer sunlight, which has a maximum erythemal irradiance under clear skies around midday of 7–9 SED h^{-1}. This means that whilst it takes about 20–25 min of sun exposure on an unshaded, horizontal surface to receive sufficient solar UV to elicit a minimal erythema in unacclimatised white skin a few hours later, the same reaction can be achieved with the UV solar simulator in about two minutes. The total unweighted irradiance from the UV solar simulator is about 0.35 kW m^{-2}, which is around one-third of the full spectrum irradiance from sunlight.

The general procedure for SPF determination is as follows.

On Day 1, rectangular test sites that are free of blemishes are located on the mid-back of the volunteer, on either side of the mid-line. Within each site, five sub-sites are required for each UV exposure. A series of five UV exposures, increasing in 25% increments, is administered to each sub-site (figure 7.4).

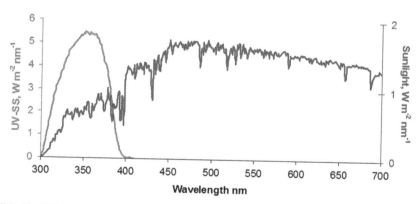

Figure 7.3. The UV spectral irradiance of clear sky, midday summer sunlight at mid-latitudes (blue) and an Oriel® Sol-UV Series UV Solar Simulator (red).

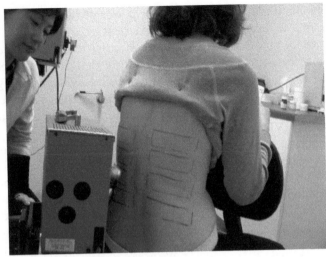

Figure 7.4. A volunteer undergoing irradiation with a UV solar simulator in order to determine the SPF of a sunscreen product (courtesy of Dr John Staton, Dermatest Pty Ltd, Australia).

On Day 2, the subject returns to the testing laboratory within 16 to 24 h after completion of the initial UV exposures to determine their erythemal sensitivity on unprotected skin. This is achieved by determining the minimal erythema dose (MED), which, in the case of sunscreen testing, is defined as the smallest UV exposure that produces just perceptible redness of the skin with clearly defined borders.

The test product, as well as standard reference product of known SPF, are applied at a surface density of 2 mg cm^{-2}. To aid uniform coverage, droplets of the product are deposited with a syringe or pipette and then spread over the test site with light pressure using a finger cot. This amount is selected to ensure even application of the product over the test area as experience shows that achieving a uniform layer with a lower surface density is difficult even though typical usage is closer to an average application thickness of 1 mg cm^{-2} (see section 7.7.1).

Sunscreen is normally allowed to dry for 15 to 30 min before commencing a series of five progressively increasing UV exposures administered to the sites treated with the test product and standard. The exposure series is determined by the product of the expected SPF of each test product and the volunteer's initial unprotected MED. For the sunscreen-protected sites, the centre of the UV dose range is that of the unprotected MED multiplied by the expected SPF of the product.

Finally, the volunteer returns on Day 3 for evaluation of responses. SPFs are computed for both the test product and standard sunscreen according to the MED on sunscreen protected skin divided by the MED on unprotected skin. Clearly, the SPF determined on the standard product needs to be within defined limits for the test procedure and results to be regarded as valid.

7.4 *In vitro* assessment of protection factors

The principle of *in vitro* assay is to measure the transmission of UV radiation through a UV-transparent substrate, apply the sunscreen, and repeat the transmission measurement.

Commercial instruments, such as the Labsphere UV-2000S transmittance analyser (North Sutton, NH, USA), are available to determine the diffuse transmission spectrum of UV radiation through the substrate before and after application of the sunscreen.

The ratio of the transmission on a wavelength-by-wavelength basis without and with the sunscreen applied gives the monochromatic protection factors at each wavelength. An index of protection e.g. SPF, critical wavelength, UVA protection factor (UVA-PF), is obtained by appropriate mathematical manipulation of the monochromatic protection factors, sometimes in conjunction with a reference solar UV spectrum and biological action spectrum.

A wide range of substrates have been used for *in vitro* assay. These include wool, pig skin, lyophilised pig epidermis, hairless mouse epidermis, human epidermis, human stratum corneum, synthetic skin casts, surgical tape, a combination of a biomembrane barrier with a biomacromolecular matrix, and rough, moulded Poly (methyl methacrylate) plates.

7.5 What wavelengths should sunscreens protect against?

Whilst our photobiological knowledge continues to increase, we still remain largely ignorant of the action spectra for the induction of basal cell carcinoma, malignant melanoma and immunosuppression, although in the case of melanoma it is sometimes acknowledged that UVA may play a larger role than it does for erythema and squamous cell carcinoma. Consequently, the logical approach to sun protection is to seek to reduce total sunlight exposure rather than selectively modify the spectrum of UV absorbed by the skin. It would seem sensible, therefore, to use sunscreens that absorb uniformly throughout the UV spectrum; that is, absorb both UVB and UVA to much the same extent. A sunscreen that exhibits more or less equal absorption at all wavelengths across the UV waveband is termed *broad-spectrum*.

The rationale behind this thinking is that our skin has evolved to exist in harmony with the mix of wavelengths that make up the spectrum of sunlight on the Earth's surface. When we seek natural shade or wear clothing to protect our skin from UV, the consequence is that we reduce the quantity (or intensity) of sunlight on our skin but change only minimally the quality (or spectrum), as illustrated in figures 6.2 and 6.10.

The benefit of using a broad-spectrum sunscreen is that for a given time in the Sun, the total UV dose absorbed by the skin is much less than if, say, a sunscreen of the same SPF but absorbing mainly UVB radiation is used.

This concept can be illustrated by figure 7.5, which shows the spectral absorption profile of three SPF30 sunscreens, but with different active ingredients, as given in table 7.2. Note that the absorption of all products falls rapidly at wavelengths in the visible region otherwise they would look coloured on the skin.

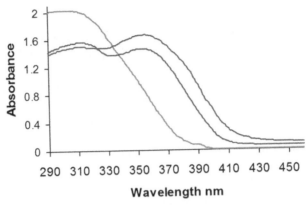

Figure 7.5. The spectral absorbance of three SPF30 sunscreen products: Product A—red curve; Product B—green curve; Product C—blue curve.

Product A contains active ingredients that absorb mainly UVB radiation and is typical of sunscreens used in the 1980s and 1990s.

Product B combines active filters that absorb both UVB and, to a lesser extent, UVA and typifies many products that have been on the market for the past 20 years or so. And product C is a modern, broad-spectrum sunscreen that provides balanced protection across the UV spectrum.

Another way of thinking about these three different classes of sunscreen would be to regard A-type products as providing primarily UVB protection, B-type products as providing partial broad-spectrum protection and C-type products as providing optimal broad-spectrum protection.

Suppose that three men, Arthur, Bill and Charles, use products A, B and C, respectively. Each product is formulated to result in a labelled SPF of 30. They each apply the respective products to their skin at a thickness of 2 mg cm^{-2} and are exposed to the Sun in exactly the same way and for the same time. However, the UV dose on the skin received by each subject will differ. Bill will receive a UV dose that is twice that received by Charles, whilst Arthur receives 6–7 times the UV dose to his skin compared with Charles. It is clear, therefore, that broad-spectrum products minimise the overall UV exposure of skin for a given SPF.

It may be tempting to believe that it is beneficial to increase the absorption of sunscreens in the UVB region relative to the UVA to reflect the fact that skin damage is associated more with UVB than UVA exposure, as discussed in chapter 4. However, this belief is a fallacy since a sunscreen exhibiting higher absorption in the UVB waveband compared with the UVA, as exemplified by Product A in figure 7.5, confers no benefit in terms of erythema (and endpoints with similar action spectra) than a sunscreen with the *same* SPF that exhibits uniform absorption at all wavelengths throughout the UV spectrum. More importantly, the spectral profile of Product A, and to a lesser extent Product B, offer inferior protection when endpoints with other action spectra are considered, as well as resulting in a total UV burden to the skin that is higher than sunscreen products showing a uniform spectral absorption profile, as illustrated by the above example.

Table 7.2. The active ingredients in each sunscreen whose spectral absorbance profile is shown in figure 7.5.

Product	Active ingredients
A	Benzophenone-3, ethylhexyl methoxycinnamate, phenylbenzimidazole sulphonic acid
B	Bis-Ethylhexyloxyphenol methoxyphenyl triazine, diethylamino hydroxybenzoyl hexyl, ethylhexyl methoxycinnamate, ethylhexyl triazone
C	Bis-Ethylhexyloxyphenol methoxyphenyl triazine, diethylamino hydroxybenzoyl hexyl benzoate, ethylhexyl methoxycinnamate, methylene bis-benzotriazolyl tetramethylbutylphenol, octocrylene

It is now widely accepted that sunscreens should provide balanced, broad-spectrum protection—sometimes referred to as *spectral homeostasis*—and so manufacturers have sought to develop products with spectral profiles that approach this ideal. There has been considerable success in this regard and today we have available products that virtually meet this criterion.

In order for consumers to make informed choices about sunscreens, metrics are necessary that convey both the amplitude and breadth of protection. The former is expressed by the SPF, but as yet there is no universal agreement on how to express the broad-spectrum property of a sunscreen, as we shall see in the following section.

7.6 Broad-spectrum protection

There are a number of *in vivo*, *ex vivo*, and *in vitro* methods that have been described to measure broad-spectrum, or UVA, protection of sunscreens, which include:

In vivo
Sensitise skin to UVA by using photoactive agents (e.g. 8-methoxy-psoralen).
Erythema in unsensitised skin induced by a UVA-emitting source.
Persistent pigment darkening (PPD).
Diffuse reflectance spectroscopy.

Ex vivo
Spectrophotophometry using excised human or mouse epidermis as a substrate.

In vitro
Dilute solution/thin film.
Spectrophotophometry on an artificial substrate.

Whilst *in vivo* SPF determination remains the criterion standard for the magnitude of protection, methods for the assessment of broad-spectrum (UVA) protection have been a matter of debate for more than 20 years. A consensus of sorts has been reached with *in vitro* measurements based on substrate spectrophotometry generally being preferred to *in vivo* or *ex vivo* methods.

7.6.1 The current situation regarding broad-spectrum protection

Regrettably we presently have a range of metrics determined from *in vitro* spectrophotometry that have been proposed to express broad-spectrum protection and as of now there are differing recommendations from a European and North American perspective.

In its final rule published in June 2011 [7], the Food and Drug Administration adopted a pass/fail test using the *in vitro* critical wavelength as the sole method for assessing UVA or broad-spectrum protection. The critical wavelength method [8] evaluates the wavelength at which 90% of the cumulative area under the total absorbance curve from 290–400 nm occurs. The FDA specifies that only products with a critical wavelength equal to, or greater than, 370 nm can be labelled as broad spectrum.

The European Commission has recommended a minimum ratio of UVA-PF to SPF of at least one-third for all sunscreen products marketed in the European Union [9]. The UVA-PF is based on determination of a UVA protection factor obtained by application of the persistent pigment darkening method [10] and is determined according to ISO standard 24443 [11]. Products that meet or exceed this criterion are permitted to display the symbol shown in figure 7.6(a) on their packs.

The so-called Boots method that is used in the UK is based on a rating given as the ratio of mean UVA (320–400 nm) absorbance to mean UVB (290–320 nm) absorbance. This ratio was introduced as the Boots star rating in the UK in January 1992 and modified in March 2008. The star rating system is owned and licensed for use by The Boots Company PLC. Only one of the five logos shown in figure 7.6(b) would appear on the reverse of the bottle depending upon the numerical value of the UVA:UVB mean absorbance ratio. However, sunscreen performance has improved appreciably since the star rating was introduced such that 1- and 2-star ratings are no longer needed.

Yet all these *in vitro* indices are different mathematical manipulations of the same data set; that is, the spectral absorbance of a layer of sunscreen applied to an artificial substrate and one metric is no more inherently accurate than any other. *In vitro* sunscreen assay conforms to the first uncertainty principle in classical mechanics, which is that information cannot be created; only measurement, not computation, can reduce uncertainty.

The situation that now prevails is not in the interest of consumers, who simply do not understand why there need to be different measures and expressions of how well a sunscreen protects against UV radiation. Consumers do not want to make choices about something they do not understand but fear may nonetheless be important to their health. Studies in other situations, particularly with workers in the nuclear industry, suggest that non-specialists would rather trust the organisation to take

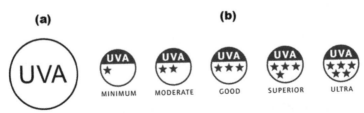

Figure 7.6. Symbols used on European sunscreens that indicate broad spectrum protection: (a) symbol approved by the European Commission, (b) Boots star rating logo. Courtesy of Walgreens Boots Alliance.

whatever steps the professionals deem to be appropriate with regard to health, rather than be confronted with technical choices, which many find bewildering.

This does not imply that consumers are not concerned about the efficacy of products they apply to their skin. They are, but the relevant environment for them to monitor, and the information to gather, evaluate and interpret is mainly social knowledge pertaining to the trustworthiness of the manufacturers of sunscreens on which they know they are unavoidably dependent.

7.7 The sunscreen–sunburn paradox

A paradoxical result of many observational studies is the high prevalence of sunburn in subjects using sunscreen. So why might this be given that the prevention of sunburn is by far the most common reason for using sunscreens and laboratory testing confirms unequivocally that sunscreens will prevent erythema?

How large the SPF should be to prevent sunburn can easily be determined if we know the local solar UV levels, the time spent and the behaviour outdoors of someone using sunscreen, and their personal susceptibility to sunburn. Maximum ambient daily UV levels, which occur on the ground, under a completely open (i.e. no shade), cloudless summer sky range from about 40–70 SED over the latitude range 10° to 60° N (table 5.5)

These maximum ambient exposures will not be received by people simply because it would be unrealistic to lie flat in the unshaded sun all-day without moving. An extreme sunbather might spend half the time lying on their back and half the time supine, resulting in a maximum exposure on much of the body surface of about 50% of ambient. For people who are upright and mobile engaging in an outdoor pursuit such as gardening, walking or sport, the exposure relative to ambient on commonly exposed sites, e.g., chest, shoulder, face, forearms and lower legs, ranges from about 5% to 60%, depending on the particular body site and the presence of nearby shade, such as buildings.

So someone in southern Europe who spends all day long sunbathing on a clear summer's day would receive a daily skin exposure of no more than 60 SED × 50% = 30 SED. If they have skin that responds typically to sunlight, they would need to use a sunscreen that reduces this exposure to, say, 3 SED or less if they are to avoid sunburn that evening. In other words, the sunscreen should provide 30 SED divided by 3 SED, that is ten-fold protection, or put another way, have an SPF of 10. And if you are walking around in an urban environment, where you may receive about 20% of ambient on exposed sites such as the arms and face, sunburn should theoretically be avoided if you are wearing a product providing an SPF of [60 SED × 20%]/3 SED = SPF 4.

So then, why do people who use high factor (SPF > 15) sunscreen experience sunburn so frequently? That the protection achieved is often less than that expected is explained by one or more of the following factors.

7.7.1 Application thickness

The protection offered by a sunscreen—defined by its SPF—is assessed after testing a panel of volunteers in the laboratory at an internationally-agreed application thickness of 2 mg cm^{-2}. Yet there are many studies that show that consumers in real-life situations apply much less than this—typically between 0.5 to 1.5 mg cm^{-2}.

The relationship between product thickness and resulting *effective* SPF has been investigated by a number of authors. The results have been variable ranging from an exponential relationship between SPF and applied thickness through to an approximately linear relationship, with others showing a relationship somewhere between exponential and linear. It would be fair to say, however, that for sunscreens with spectral profiles typical of those currently available, a linear relationship is closest to the best descriptor and so most users are probably achieving a level of protection of between one-quarter to three-quarters of that expected from the SPF on the product label.

7.7.2 Application technique

When sunscreens are tested on a panel of volunteers in the laboratory to determine their SPF, great care is taken to achieve a uniform layer of sunscreen over the test area by spreading with a gloved finger. In practice, of course, nothing like this care is taken when sunscreen is applied to the skin. Sunscreen is normally spread haphazardly and non-uniformly with the result that patchy sunburn may appear after sun exposure, as illustrated in figure 7.7.

Even when people decide to use sunscreen, they may not always choose to apply it to all exposed sites. Areas most likely to be missed during sunscreen application are the ears, neck, feet and legs.

7.7.3 Sunscreen type

Sunscreens containing inorganic chemicals, such as zinc oxide or titanium dioxide, as the sole active ingredient can leave a white film on the skin and, as a consequence, people may compensate for this by applying less quantity than products incorporating organic filters. This can reduce the protection achieved, which is especially unfortunate because inorganic sunscreens will tend to be preferred by sun-sensitive individuals, including those with sensitive skin or eczema.

7.7.4 Sunscreen formulation

The formulation of the sunscreen can be an important factor influencing an individual's willingness to use and reapply a sunscreen. In terms of formulation, creams have lost popularity due to the fact that they are less easy to rub into the skin and can leave a white film, which many consumers find unacceptable. High-SPF products, which could previously be found only in cream formulations, now include lotions, milks, gels, sticks, and sprays.

The importance of good formulation, resulting in a cosmetically-appealing product, is increasingly being recognised by manufacturers. It is all very well making a sunscreen with a very high SPF but if consumers do not like the feel or look of it, they will not bother using it, or may apply so little that the real protection they are getting leaves them vulnerable to sunburn and skin damage. Figure 7.8

Figure 7.7. The consequence of non-uniform sunscreen application (photo courtesy of http://www.philip.greenspun.com).

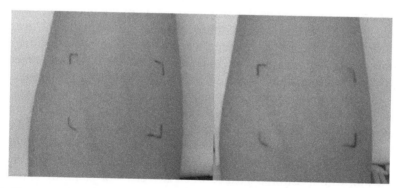

Figure 7.8. The sunscreen on the left leaves an invisible film on the skin, while that on the right leaves a white residue (courtesy of Dr Marc Pissavini).

compares two sunscreens, one with a formulation that results in a colourless film on the skin and the other that tends to leave a white residue.

7.7.5 Water immersion

Many people take part in activities such as swimming, bathing and water sports while wearing sunscreens and this has led to the development of products that bind well to the skin surface and resist wash-off, so much so that the majority of sunscreen products now available for recreational use claim to be water-resistant.

7.7.6 Reapplication

The reasons for reapplying sunscreen during sun exposure are to compensate for initial under-application so as to achieve an SPF more in line with the rated value, and to replace sunscreen that may have been removed by water, vigorous towelling, or friction with clothing or sand. Inadequate application is the primary purpose for reapplication since modern water-resistant sunscreens adhere well to the surface of the skin even after immersion in water.

Guidance on sunscreen packs about reapplication is generally to 'reapply frequently' or 'reapply regularly' and a common recommendation by many public health agencies is to reapply sunscreen every 2–3 h. Given considerations of cost, convenience and human nature, it is unlikely many users will reapply sunscreens frequently or regularly.

If you use a sunscreen that is easily removed from the skin then you will achieve little in the way of sun protection, no matter when it is reapplied. For water-resistant products, the lowest skin exposure results from early reapplication into the sun exposure period, and not at 2–3 h, after initial application. Typically, reapplication of sunscreen at 20 min into a 4 h exposure period to strong summer sunshine results in one-half to three-quarters of the UV exposure that would be received compared with waiting 2 h before the sunscreen is reapplied.

Consumers are advised to apply sunscreen liberally or generously. That people generally apply less than 2 mg cm^{-2}—the thickness used by manufacturers during the testing process to determine the SPF—has led many commentators into the trap of believing that consumers use inadequate amounts of sunscreen for protection. The reality is the reverse. People use the quantity they feel comfortable with and in this sense are using the 'correct' amount; it is the labelled SPFs that are misleading.

It is likely, therefore, that campaigns to encourage people to apply a greater quantity of sunscreen at a single application will fail. An important factor in sunscreen performance is the uneven nature of the skin surface. If an opaque sunscreen is applied to the skin, the surface markings become visible as the grooves on the skin surface are filled. With further application, the intervening ridges are also covered and the surface becomes more featureless. The situation is analogous to painting a wall with a textured surface when two coats of paint are almost always required for satisfactory coverage. In the same way two 'coats of sunscreen' may be required for adequate protection.

Although consumers generally apply insufficient sunscreen product to achieve the labelled SPF, this does not mean the product failed or there is no efficacy. On the contrary, there are considerable data that support the view that protection from solar UV is achieved under 'normal' use conditions.

Finally, people will tend to apply sunscreen more frequently on those summer days when the weather is fine and they intend spending recreational time outdoors. And it is on days such as these that they are most vulnerable to sunburn if sunscreen application is less than ideal.

7.8 SPFs in natural sunlight

In the previous section we saw how the use and application of sunscreen, coupled with removal due to factors such as sweating and rubbing during sun exposure, generally results in lower protection than indicated by the labelled SPF. Another factor that contributes to this mismatch between expected and delivered protection is the difference between the spectral emission of UV solar simulators that are used in the laboratory determination of SPF and the spectral power of terrestrial sunlight.

There are a few reports indicating that SPFs determined using UV solar simulating radiation in the laboratory are higher than those measured using natural sunlight as the source, but the data obtained with sunlight, especially on modern sunscreens with SPFS of 15 and higher, are inadequate to give a clear indication of the magnitude of the difference.

This is not surprising since determining SPFs using natural sunlight is problematic. With high factor sunscreens, in particular, the exposure periods required would be an hour or more. During this time not only does the spectral irradiance of sunlight change on a minute-to-minute basis as the solar altitude, and possibly cloud cover, changes, but perturbations in air temperature, humidity and subject movement, for example, will also act as confounding factors. So while outdoor testing under actual use conditions might be considered the most realistic method, it is both impractical and unreliable.

The spectral emission of a typical UV solar simulator used for assaying the protection provided by sunscreens is compared in figure 7.3 with the spectrum of summer sunlight where it is obvious that, unlike sunlight, a solar simulator emits very little radiation beyond 400 nm.

In order to know whether the removal of visible light from UV solar simulators is an issue, we need to consider the erythema response of the skin to different wavelengths of UV and visible radiation.

The erythema action spectrum shown by the red curve in figure 4.9 was determined up to 405 nm and despite irradiation at 435 nm, erythema was not seen. This does not mean that the skin will not develop erythema in response to visible light, simply that the exposure dose given was insufficient to induce a reaction.

We postulate that erythema induction does not cease at around 400 nm but continues at wavelengths into the visible region (see section 4.2.6), albeit with an ever-decreasing effectiveness as wavelength increases. As there are no data on the action spectrum beyond 405 nm, the best we can do is to extrapolate the curve on the same downward slope between 380 and 405 nm.

The SPF of a sunscreen is defined by:

$$SPF = \sum_{290}^{\lambda_{max}} E(\lambda)S(\lambda)\Delta\lambda \bigg/ \sum_{290}^{\lambda_{max}} E(\lambda)S(\lambda)T(\lambda)\Delta\lambda$$

$E(\lambda)$ is the spectral power distribution of the radiation source used in the determination, and $S(\lambda)$ is the erythema action spectrum shown by the red curve in figure 4.9, together with the presumed action spectrum in the visible region extrapolated to wavelengths beyond 405 nm as explained above. $T(\lambda)$ is the spectral

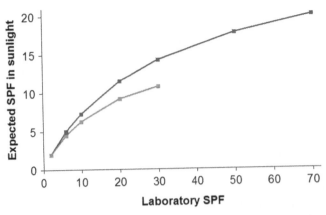

Figure 7.9. The variation of SPF expected in natural sunlight with laboratory determined SPF using a UV solar simulator for Product A (red curve) and Product C (blue curve).

transmission through the sunscreen and is calculated from the sunscreen absorbance spectrum, $A(\lambda)$, as $10^{-A(\lambda)}$. The upper limit of the summation, λ_{max}, is, theoretically, the maximum wavelength in terrestrial sunlight, but it can be truncated at 700 nm without affecting the accuracy of the SPF determination.

We consider two of the sunscreens described in table 7.2; Product A, which is typical of sunscreens used in the 1980s and 1990s, and Product C, a modern, broad-spectrum sunscreen that provides balanced protection across the UV spectrum. The spectral absorption profile of both these products, scaled to give identical SPFs of 30 calculated using the above equation when the radiation source is a UV solar simulator, are shown in figure 7.5.

If now, we use the spectrum of natural sunlight (blue curve, figure 7.3) in the equation we find the SPF has fallen from 30 to 15, since natural sunlight has much greater output in the long wave UV and visible regions compared with the UV solar simulator, and this compensates to some extent for the falling erythemal sensitivity of skin in this spectral region.

By scaling the absorption profiles appropriately, we can estimate SPFs determined with the UV solar simulator and the corresponding SPF expected in sunlight (figure 7.9).

Note that because the absorption of the active UV filters in Product A are limited in the UVA region, it is not possible to achieve SPFs higher than 30 at allowable concentrations of the active ingredients used in these products.

We see from figure 7.9 that the mismatch between the predicted laboratory determination of SPF using a UV solar simulator and that expected in natural sunlight increases as the SPF increases such that for products labelled SPF50+, it is not possible to achieve a protection against solar radiation of much more than 25-fold. Also, for a given SPF, the degree of mismatch is greater for sunscreens, such as Product A, that provide sub-optimal broad spectrum protection.

Another observation from figure 7.9 is the non-linear relationship between laboratory (or labelled) SPFs and the SPF expected in sunlight. For example, a broad spectrum sunscreen, like Product C, may be available as SPF15 and SPF30, i.e. a factor of two difference in protection. But the expected SPFs in sunlight for these two products are SPF10 and SPF15, respectively, i.e. a factor of 1.5 difference in protection.

We conclude, therefore, that labelled SPFs do not reflect the magnitude of protection expected in sunlight and that the interpretation of the SPF, often found in beauty and other magazines, that it can be thought of as how much longer skin covered with sunscreen takes to burn compared with unprotected skin, can no longer be sustained.

7.9 Compliance

It is clear that the UV exposure of sunscreen-protected skin depends not just on the absorption characteristics of the product but also on a number of other factors to do with application, such as how much and how well sunscreen is applied. Consequently, the efficacy of a product can be thought of as the combination of the technical performance of its active UV filters and how pleasing the sunscreen is to apply and use.

One consequence of the trend for higher SPFs is that, in general, the water content of the formulation decreases as the concentration of UV filters increases to deliver the higher protection, with the result that in many cases the product is more difficult to spread and users compensate by applying a smaller amount than they might do with lower SPF products.

Good aesthetic properties (appearance and skin feel) are now seen as an important aspect of sunscreen products, as they encourage greater consumer compliance. Studies with a panel of volunteers indicate that what is usually desired is a product that spreads easily with a moderately wet feeling during application but feels smooth and dry afterwards with little or no perceivable residue. Consequently, those products with good sensory properties are associated with a subjective assessment of a pleasurable product, which results in a higher application thicknesses and hence greater delivered photoprotection than products with negative attributes such as greasiness. These steps, from the expectation to realisation of the photoprotection delivered by a suncare product, are illustrated diagrammatically in figure 7.10 and confirm that the sun protection benefit associated with sunscreen use is strongly linked to cosmetic attributes and compliance.

As manufacturers strive for greater photoprotection in their products, this needs to be accompanied by close attention to cosmetic (or galenic) attributes so that the improved UV-absorbing properties of the products are translated to the user. What makes the product pleasing to use (compliance) is as, or more, important as the technical benefit the product delivers (performance).

7.10 Impact of sunscreen SPF on the likelihood of sunburn

In January 2011, the National Institute for Health and Clinical Excellence (NICE) in the UK issued a public health guidance on skin cancer prevention and recommended that sunscreens possessing a sun protection factor (SPF) of at least

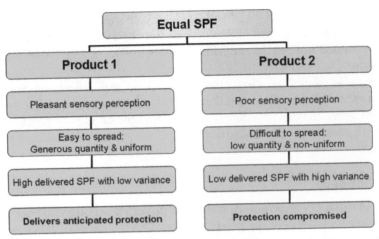

Figure 7.10. The steps from expectation to realisation of sunscreen efficacy for two different products having the same labelled SPF.

15 were sufficient if applied adequately [12]. More recently, in June 2011, the US Food and Drug Administration gave similar advice [7].

In the introduction to section 7.7, it was argued that a photoprotective device, whether a sunscreen, shirt or shade, need only possess a protection factor of 10 to prevent sunburn during all-day exposure to Mediterranean summer sunshine, so using SPF15 should be adequate if sunburn is to be avoided.

Yet from the discussion that followed, it was shown why sunburn may occur with sunscreens labelled SPF15 or higher, due to factors such as low application thickness, non-uniform spreading of the product on the skin, and overestimation of labelled SPFs compared with those expected in natural sunlight.

7.10.1 UV exposure on a sun-seeking holiday

Consider someone on a sun-seeking holiday in an area of high insolation, such as Florida (latitude 25–30°N) where the maximum ambient, clear sky erythemal UV during a summer's day is about 70 SED (table 5.5). Let us assume that this holidaymaker is outside for 6 h around the middle of the day; during this time, the ambient UV exposure would be about two-thirds of the daily total, or around 45 SED.

Assume also that the subject is exposed in a largely unshaded environment, such as at the beach, and spends a mixture of time both ambulant and prone/supine, and so we take a representative value of 0.4 as the fraction of ambient UV received on exposed skin. Hence, his/her exposure during 6 h of summer sun-seeking behaviour might typically be 45 × 0.4, or 18 SED.

We saw in section 4.2.2 that it requires an exposure of about 2–5 SED to result in minimal erythema on unacclimatised white skin (figure 4.8). If we take an exposure of 6 SED then any area of skin not protected by sunscreen by a factor of at least 3 (i.e. 18 SED divided by 6 SED) would probably result in sunburn.

7.10.2 Estimating the fraction of sunburnt skin

Manual application of a topical sunscreen, coupled with an irregular skin surface topology, makes it impossible to achieve a uniform film thickness. Experimental studies indicate that the distribution of product thickness applied to the skin surface is not uniform but exhibits a positive skew, the degree of skewness reflecting the formulation of the product. This means that even though the average thickness may be around 1 mg cm^{-2}, for example, there will be some areas of skin where there is little or no protection from the applied product, which can result in sunburn as illustrated in figure 7.7, and others where the local thickness may be in excess of 2 mg cm^{-2}.

Various mathematical approaches have been investigated to model the distribution of film thickness across the skin surface and the one that appears to give the best fit to experimental data, when sunscreen is applied on a skin substrate and investigated using topographical measurements, is a gamma function.

It follows that the fraction of skin area to which sunscreen of nominal SPF (SPF$_{labelled}$) is applied at an average thickness of X mg cm^{-2} and that is protected at or below a given threshold SPF (SPF$_{threshold}$) is expressed as:

$$\text{GAMMADIST}(2 \times (\text{SPF}_{threshold} - 1)/(\text{SPF}_{labelled} - 1), \alpha, X/\alpha, \text{TRUE})$$

GAMMADIST is an Excel function that that returns the cumulative gamma distribution and α is the gamma shape factor that expresses the asymmetry of the distribution; the smaller the value of α, the greater the skewness and unevenness of the thickness distribution.

In this simulation exercise, we take a mean application thickness of 1 mg cm^{-2} and assume a value for the gamma distribution shape factor (α) of 1.4. This value equates to a ratio of the mean to median application thickness of 1.3, which is typical for an oil-in-water recreational suncream [13].

From the exposure scenario outlined in section 7.10.1, we require a protection of at least three-fold over the skin to which sunscreen has been applied to prevent sunburn, since any area of skin not protected to this level is likely to show definite to marked erythema. A three-fold protection in sunlight requires an SPF$_{threshold}$ of 3.5 determined in the laboratory (figure 7.9).

Evaluating the expression above we find that for sunscreens labelled SPF15 and SPF30, the percentage of skin area that is protected three-fold or less in sunlight is approximately 23% and 10%, respectively. This assumes that the sunscreen binds well to the skin and the surface density remains unchanged throughout sun exposure, as would be roughly expected from products that claim to be water resistant.

While either sunscreen, if delivering their nominal SPF over the entire exposed skin, would be sufficient to prevent any erythema, the simulation indicates that the combination of the average quantity applied with the variability in thickness over the skin surface will lead to patchy erythema. Even though there is only a factor of two in labelled SPFs, the extent of erythematous areas of skin in SPF15 sunscreen users is greater than two-fold compared with those using SPF30 sunscreen.

People who intend spending long periods outside in strong sunshine would be better advised to use sunscreens labelled SPF30 (or higher) rather than SPF15 sunscreens if they wish to minimise their risk of sunburn. Equally important is to choose a sunscreen with good cosmetic properties, which will allow the product to be spread as homogeneously as possible.

The corollary to this advice is that sunscreen use does not deliberately lead to prolonged sun exposure as inadequate and non-uniform application could result in unacceptably high exposure of partially protected areas of skin that may result in marked, or even painful, sunburn.

7.11 Safety of sunscreens

Among the human safety concerns related to sunscreen use are inhalation of droplets from using spray products, vitamin D deficiency, endocrine disruption, systemic absorption from long-term use, and the risk to humans from the use of nano-structured titanium dioxide or zinc oxide in sunscreen products.

Expert reviews of sunscreen safety and efficacy conclude that the current list of commonly-used organic and inorganic active ingredients do not pose a concern for human health [14, 15]. However, although uncommon, sunscreens can occasionally cause skin irritation, or a skin allergy. Nevertheless, concern about any adverse effects of using sunscreen continues to generate public interest.

7.12 Shelf life of sunscreens

Sunscreens are emulsions of oil and water and so will always tend towards separation. The time this takes depends on the quality of the formulation and can vary from just a few months to many years. Typically, the shelf life for sunscreens is around 30 months.

To maximise their quality, sunscreens should be stored in a cool, dry place, out of direct sunlight. The smell and feel of the product may give an indication about whether it is fit to use, but as products have a fixed life after they have been opened, reference should be made to the 'period after opening' symbol on the labelling.

Finally, there is the question of microbiological integrity once the sunscreen has been opened and used for the first time. This tends to shorten physical stability, particularly these days when allergy concerns lead to the minimal use of preservatives in sunscreens.

7.13 Do sunscreens prevent skin cancer?

Consumer knowledge of the harmful effects of sunlight has increased dramatically in the past two decades, largely due to the combined efforts of public health agencies and the media. People are now much more aware of the risk of skin cancer–the most common human cancer with over 2 million people worldwide each year getting skin cancer—from too much sun exposure and will apply sunscreens in the belief that this risk can be reduced by their use. So what evidence is there that sunscreens are effective in this important public health arena?

7.13.1 Sunscreen use and prevention of non-melanoma skin cancer

The strongest available evidence that sunscreen use is an effective approach to prevention of basal cell cancer (BCC) and squamous cell cancer (SCC), comes from the results of a community-based trial in which 1621 adults aged 25 to 75 were randomly selected from all residents of Nambour, a subtropical Queensland township [16].

Trial participants were randomised either to apply a freely supplied sunscreen with an SPF of 16 daily to the head and arms for 4½ years, or to continue their usual level of sunscreen use or non-use. In comparison with people not asked to use sunscreen on a daily basis, the intervention group showed a 40% reduction in SCC tumours at the conclusion of the trial. Eight years after cessation of the sunscreen intervention trial, participants who had been randomised to daily sunscreen use continued to show a 40% decrease in SCC incidence.

Although there was no effect on BCC incidence during the trial period, there was a trend of increasing intervals between BCCs among the treatment group compared with people in the control group who developed multiple BCCs.

7.13.2 Sunscreen use and prevention of melanoma

The observation that sunscreens protect against sunburn led to the common expectation that they will also protect against skin cancer, including malignant melanoma. However, it was not until the results of 15 case-control studies were reviewed by an expert panel in 2000 to evaluate the potential preventive effect of sunscreens against malignant melanoma was there some cause for concern [17]. Of the 15 studies examined, four provided little evidence of an effect of sunscreen use on the risk of melanoma, three studies showed significantly lower risks for melanoma in sunscreens users compared with non-users, while the remaining eight studies showed significantly higher risks in sunscreen users.

More recent analyses of case–control studies have demonstrated no association between sunscreen use and either the prevention or development of malignant melanoma. The inconclusive data concerning the role of sunscreens in preventing melanoma is, in part, due to both positive confounding, e.g. people who are at most risk of burning and most likely to develop melanoma are also most likely to use sunscreens, and negative confounding, e.g. sunscreen users may also use other methods of sun protection such as clothing.

Yet if consideration is given to the period (1970s to 1990s) during which the data used in most case–control studies were collected, the UV absorbing properties of sunscreens prevalent at that time, and how sunscreen is used and applied in practice, then the observation that sunscreens appear to play little or no role in preventing melanoma is not unexpected as these products probably resulted in an effective SPF of only 2–3, with little or no protection against UVA wavelengths.

On the other hand, a modern, broad-spectrum sunscreen labelled SPF30 to SPF50 applied at an average thickness of around 1 mg cm^{-2}, will result in an effective SPF in sunlight of around 10 and a total solar UV radiation dose to the skin of around one-quarter or less than that of earlier generation products.

To date, there are two prospective studies that suggest that sunscreens may have a role to play in preventing melanoma. Ten years after cessation of the sunscreen intervention trial in Nambour referred to above in section 7.13.1, the group of 812 people randomised to daily sunscreen use had experienced 11 new melanomas, compared with the control group of 809, who experienced 22 new melanomas [18]. While the number of melanomas in the daily sunscreen group was one-half of the number appearing in the control group, the small number of melanomas that actually appeared in both groups meant that the result is of borderline statistical significance and so we still await confirmatory evidence that sunscreens really do prevent melanoma.

Secondly, a prospective population-based study in Norway [19] of 143 844 women aged 40 to 75 years in which 722 cases of melanoma were observed over a ten-year period, concluded that during intentional sunbathing, use of SPF15+ sunscreen can reduce melanoma risk compared with use of SPF <15 sunscreen. Moreover, the authors suggested that use of SPF15+ sunscreen by all women in the age range 40 to 75 years could lead to an 18% drop in melanoma incidence in approximately 10 years.

7.13.3 Sunscreens, melanoma and the precautionary principle

We have, then, a dilemma. Applying the principles of evidence-based medicine there is not the strength of evidence to use sunscreens as a preventative measure in melanoma as would be expected before a new drug was introduced as a therapeutic intervention.

However, while evidence-based medicine works well when applied to therapeutic interventions, it is not necessarily effective when it comes to the prevention of diseases such as cancer, especially as there are very few studies demonstrating that sunscreens have a protective effect against melanoma.

To acquire the evidence about whether or not modern, high SPF broad-spectrum sunscreens are effective in reducing the risk of melanoma would take a decade or more, or more likely, may never become available owing to the complexity of conducting prospective trials in cancer incidence and human behaviour. Because of this lack of evidence, some would argue that the focus of recommendations for melanoma prevention in public health campaigns should be more emphasised to sun avoidance, shade and clothing.

On the other hand, the precautionary principle, which states that if an action or policy might cause severe or irreversible harm to the public, then in the absence of a scientific consensus that harm would not ensue, the burden of proof falls on those who would advocate taking the action. In other words, those who say that sunscreens should not be used as a preventative measure in melanoma because of lack of evidence for their efficacy must demonstrate this lack of efficacy for their advice to be followed. Logic would suggest that demonstration of lack of efficacy would be difficult as exposure to UV radiation is widely recognised as a risk factor in melanoma and modern sunscreens attenuate the intensity of solar UV entering the skin.

7.14 The population impact of sunscreen use on skin cancer incidence

As an example, we consider how the incidence of melanoma in the UK might be influenced by the regular use of sunscreens.

The potential number of melanomas prevented in the population by encouraging the use of sunscreen, especially during recreational sun exposure, may be calculated as [20]:

$$N[1/(1 - P(1 - RR)) - 1]$$

where N is the number of cases of melanoma presenting each year in the population of interest, P is the prevalence of sunscreen use and RR is the relative risk estimate for the protective effect of regular sunscreen use.

The number of cases of melanoma (N) in the UK is currently running at about 16 000 cases per year. Sunscreen is a popular form of sun protection, especially during sunbathing and at the beach. The prevalence patterns of sunscreen use vary widely between countries, gender and age but we take here a value for P of one-third (see section 7.2.4).

Currently, there are very few data that confirm the role of sunscreens in preventing melanoma (see section 7.13.2) but we use an estimate for RR of 0.5 (95% CI 0.24–1.02) taken from the only randomised trial to assess the efficacy of sunscreen on melanoma [18].

From these data we estimate that of the approximately 16 000 people in the UK who will be diagnosed with melanoma this year, 3200 cases, or 20%, would be avoided if sunscreen was used regularly by everyone rather than the presumed current prevalence of one-third.

In this analysis, we assume that regular sunscreen use is equally effective at all ages. However, it is possible that it may have different magnitudes of effect at different periods in life. For melanoma, and also BCC, UV exposure in early life is thought to be particularly important (section 4.4.6), and so sunscreen use among children and adolescents may lead to even greater benefits in cancer prevention than if used solely in later life.

It must be stressed that reliable data supporting the values for P, and especially RR, are lacking and so the estimate of 3200 cases of melanoma being prevented by widespread sunscreen use in the UK is simply illustrative and may have little basis in reality.

7.15 Sunscreens and solar infrared radiation

Reports that exposure to infrared (IR) radiation can result in deleterious effects in the skin have led to many sunscreens incorporating agents that are said to protect against IR damage.

Whilst there is experimental evidence to regard infrared exposure of the skin as a *hazard* (see section 4.5), this does not necessarily mean that IR exposure in the normal context of human behaviour outdoors is a significant *risk* to health. For harm to occur there must be both the hazard and sufficient *exposure* to that hazard.

It is estimated [21] that workers who are occupationally exposed to industrial sources of IR, such as steel and glass furnaces, will generally receive annual IR doses that exceed those received by even sun-seekers and yet chronic skin abnormalities, such as erythema *ab igne*, in these groups of workers are rare.

The rationale for incorporating agents into sunscreens that offer some protection against infrared damage must be based on a sound risk: benefit analysis. The benefit is, at present, only speculative as we lack direct observational data on the harm resulting from solar IR exposure. Risks can be both direct, largely related to the toxicological safety of the relevant agents, but also indirect; for example, would the incorporation of these agents increase the cost of products and dissuade some people from using sunscreen and so benefitting from the UV protection they undoubtedly provide?

With further studies on both the biological consequences of exposure to solar infrared radiation and confident estimates of human IR doses in a range of solar exposure scenarios, the need, or otherwise, for incorporating IR protection into topical sunscreens should become clearer.

7.16 Sunscreens and vitamin D

As discussed in section 2.3.1, vitamin D is synthesised in the skin by the action of solar UVB radiation. Since sunscreens are designed to attenuate UVB, it might be thought that applying a sunscreen would adversely affect vitamin D production.

Whilst laboratory studies have demonstrated that applying sunscreen at 2 mg cm^{-2} followed by irradiation with UV lamps can compromise vitamin D production, the same effect has not been seen in real-life field studies. There are a number of factors that can account for this.

Firstly, as we have seen in section 7.7, people generally apply less quantity of sunscreen than used by manufacturers in the testing process, and that they spread it non-uniformly, resulting in some area of skin with little or no protection. Also, there is evidence to suggest that only a minority of sunscreen users re-apply as necessary to maintain adequate sun protection.

Secondly, sunscreen is mostly used during recreational exposure when people intend remaining in the Sun for an extended period of time, so even areas of skin that are protected, may have a sufficiently thin layer of sunscreen that during the period of sun exposure, adequate UV penetrates through to the skin to synthesise clinically significant quantities of vitamin D.

And finally, vitamin D is synthesised in our skin on a daily basis, not just during recreational sun exposure, and most people living in temperate climates will not bother applying sunscreen during their adventitious exposure, such as popping out at lunchtime during the working week. So apprehension about vitamin D status should not be a reason to withhold sunscreen application during extended periods of exposure in strong sunlight.

What may be of more concern than recreational sunscreen, is the daily application of skin care products, many of which incorporate UV filters. It is possible that the protection provided by these products during adventitious sun exposure may be sufficient to compromise vitamin D production, especially in the important period of late summer and early autumn when sun exposure is important to 'top-up' our stores of vitamin D in preparation for the winter. However, there are no data to support, or otherwise, this hypothesis.

7.17 A strategy for sunscreen use

A strategy is suggested as a rational approach to the application of topical sunscreens that minimises the deleterious effects of UV exposure but that is commensurate with achieving adequate sun exposure for vitamin D synthesis and results in an annual UV burden that approaches the peak of the curve in figure 1.2.

The terms *winter* and *summer* are used in the strategy to refer to those periods when the maximum UV index is below or above 4, respectively, which roughly corresponds to:

Latitude	Northern hemisphere		Southern hemisphere	
	Winter	*Summer*	*Winter*	*Summer*
>50°	Oct–Mar	Apr–Sep	Apr–Sep	Oct–Mar
40°–50°	Nov–Feb	Mar–Oct	May–Aug	Sep–Apr
30°–40°	Dec–Jan	Feb–Nov	Jun–Jul	Aug–May
<30°	Never	All year	Never	All year

The strategy can be summarised as:
- No need for UV protection, either from recreational sunscreens or in day care products, during winter since there will be little impact on the annual UV burden.
- Daily skincare (incorporating UV filters ~SPF15) is equivocal during summer when exposure is largely adventitious, or unintentional, and of short duration e.g. weekdays when mostly indoors at work.
- Recreational sunscreen application (SPF ⩾ 30) on sunny holidays and long periods outdoors during the summer e.g. at weekends.

Adoption of this strategy should lead to the following outcomes:
- Prevention of sunburn.
- Result in about the same facial lifetime UV exposure as a 35 year old who behaves in a similar way with regard to sun exposure but who uses no sunscreen.
- Reduce the risk of skin cancer relative to a non-user of sunscreen.
- Delay the signs of photoaging.
- Ensure a moderate exposure to sunshine, especially in late summer and early autumn, to maintain acceptable vitamin D status over the winter period.

This strategy is only applicable to people with normal responses to sunlight. In people who have severe photosensitive disease, such as chronic actinic dermatitis, adherence to the principles of sun protection would apply throughout the year as this is part of their management. In addition, patients with a past history of skin cancer may wish to continue with sun protection all year even though this may not strictly be necessary.

References

[1] Osterwalder U, Sohn M and Herzog B 2014 Global state of sunscreens *Photodermatol. Photoimmunol. Photomed.* **30** 62–80

[2] FDA Department of Health and Human Services Food & Drug Administration, USA 1978 Sunscreen drug products for over the counter use: proposed safety, effectiveness and labelling conditions *Fed. Regist.* **43** 38206–69

[3] Deutsches Institut für Normung 1984 *Experimentelle dermatologische Bewertung des Erythemschutzes von externen Sonnenschutzmitteln für die menschliche Haut DIN 67501*

[4] Australian/New Zealand Standard 2012 *Sunscreen products—evaluation and classification. AS/NZS 2604:2012*

[5] COLIPA 2006 *International Sun Protection Factor (SPF) Test Method (Brussels, Belgium: European Cosmetic Toiletry, and Perfumery Association (COLIPA))*

[6] ISO 24444 2010 *Cosmetics—sun protection test methods—in vivo determination of the Sun protection factor (SPF)* (Geneva: International Organization for): Standardization)

[7] FDA Department of Health and Human Services Food & Drug Administration, USA 2011 Labeling and effectiveness testing; sunscreen drug products for over-the-counter human use *Fed. Regist.* **76** 35620–65

[8] Diffey B L 1994 A method for broad-spectrum classification of sunscreens *Int. J. Cosmet. Sci.* **16** 47–52

[9] European Commission 2006 *Recommendation on the Efficacy of Sunscreen Products and the Claims Made Relating Thereto Official J. Eur. Union* L265/39, 2006/7647/EC, 39–43, Brussels

[10] Matts P J, Alard V, Brown M W, Ferrero L, Gers-Barlag H, Issachar N, Moyal D and Wolber R 2010 The COLIPA *in vitro* UVA method: a standard and reproducible measure of sunscreen UVA protection *Int. J. Cosmet. Sci.* **32** 35–46

[11] ISO24443 2012 Determination of sunscreen UVA photoprotection *in vitro* (Geneva: International Organization for Standardization)

[12] NICE 2011 *Skin cancer: prevention using public information, sun protection resources and changes to the environment. Public health guidance 32. National Institute for Health and Clinical Excellence*

[13] Sohn M, Hêche A, Herzog B and Imanidis G 2014 Film thickness frequency distribution of different vehicles determines sunscreen efficacy *J. Biomed. Opt.* **19** 115005-1–11

[14] Nash J F 2006 Human safety and efficacy of ultraviolet filters and sunscreen products *Dermatol. Clin.* **24** 35–51

[15] Schilling K *et al* 2010 Human safety review of 'nano' titanium dioxide and zinc oxide *Photochem. Photobiol. Sci.* **9** 495–509

[16] Green A *et al* 1999 Daily sunscreen application and beta carotene supplementation in prevention of basal-cell and squamous-cell carcinomas of the skin: a randomised controlled trial *Lancet* **354** 723–9

[17] International Agency for Research on Cancer World Health Organization 2001 *IARC Handbooks of Cancer Prevention vol 5: Sunscreens* (Lyon: IARC Press)

[18] Green A C, Williams G M, Logan V and Strutton G M 2011 Reduced melanoma after regular sunscreen use: randomized trial follow-up *J. Clin. Oncol.* **29** 257–63

[19] Ghiasvand R, Weiderpass E, Green A C, Lund E and Veierød M B 2016 Sunscreen use and subsequent melanoma risk: a population-based cohort study *J. Clin. Oncol.* **34** 3976–83

[20] Olsen C M, Wilson L F, Green A C, Bain C J, Fritschi L, Neale R E and Whiteman D C 2015 Cancers in Australia attributable to exposure to solar ultraviolet radiation and prevented by regular sunscreen use *Aust. N. Z. J. Publ. Health* **39** 471–6

[21] Diffey B L and Cadars B 2016 A critical appraisal of the need for infrared radiation protection in sunscreens *Photochem. Photobiol. Sci.* **15** 361–4

Sun Protection
A risk management approach
Brian Diffey

Chapter 8

Counteract the damage resulting from solar UV radiation exposure

Once exposed to solar UV radiation, damage to vulnerable skin cells that leads to acute or chronic clinical changes will follow. The skin incorporates ways of damage limitation, either by adapting itself to reduce subsequent damage or by initiating cellular processes to repair the damage that has occurred.

8.1 Photoadaptation

Exposure of skin to solar UV radiation promotes an acute inflammatory response clinically characterised by erythema. Alongside this acute response, a number of changes occur that are adaptive, in the sense that they result in a diminished future response to equivalent doses of UV radiation. This response, described as *photo-adaptation* is not fully understood, but is viewed as comprising at least two processes: melanogenesis (tanning) and epidermal hyperplasia.

8.2 Melanogenesis

Melanin pigmentation is of two types:
- constitutive—the colour of the skin seen in different races and determined by genetic factors (see figure 4.6);
- facultative—the reversible increase in tanning in response to solar UV radiation.

Melanin is a complex polymer derived from the amino acid tyrosine. There are two major classes of melanin found in human skin:
- Eumelanin—a black-to-dark-brown insoluble material found in black hair and in the retina of the eye, and which is derived from the conversion of the amino acid tyrosine.

doi:10.1088/978-0-7503-1377-3ch8

- Phaeomelanin—a yellow-to-reddish-brown alkali-soluble material found especially in red hair, again derived from tyrosine but involving a different tyrosine–melanin pathway.

Melanin is produced in the melanocyte cells that reside in the basal layer of the epidermis. These cells synthesise melanin into cytoplasmic organelles called melanosomes that are secreted into neighbouring keratinocytes.

As shown in figure 4.4, melanin seems to function as a natural sunscreen in that it absorbs most strongly in the UVB region and diminishes as we progress through the UVA wavelengths and into the visible spectrum. The ability of melanin to protect against UV is influenced not just by the chemical make up of melanin, but by the way it is bundled into melanosomes and how the melanosomes are shaped and distributed in cells in the epidermis. In white skin, the melanosomes are ovoid with a length of about 400 nm, whereas in black skin they are about twice as long and so more effective at attenuating UV radiation.

8.2.1 Immediate pigment darkening (IPD)

This is a transient darkening of exposed skin which can be induced by UVA and visible radiation. In general, the greater the constitutive tan, the greater is the ability to exhibit IPD. Immediate pigmentation can become evident within 5–10 min of exposure to summer sun and normally fades within 1–2 h. The function of IPD is not understood.

8.2.2 Delayed tanning

The more familiar delayed tanning becomes noticeable about two to four days after sun exposure, gradually increases for several days and may persist for weeks or months. If further UV exposure is avoided, the tanned skin fades, as illustrated in figure 8.1.

Following solar UV radiation exposure, there is an increase in the number of functioning melanocytes and activity of the enzyme tyrosinase is enhanced. This leads to the formation of new melanin and hence an increase in the number of melanin granules (melanosomes) throughout the epidermis.

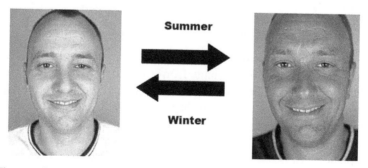

Figure 8.1. The reversible increase in tanning in response to solar UV radiation. Courtesy of The Sunday Times/News Syndication.

Although melanin shows maximal absorption in the UV region, it is not an especially effective sunscreen for white skin and it has been suggested that, contrary to popular belief, melanin is not an evolutionary adaptation to protect humans from the damaging effects of sunlight, but rather that hominids developed melanin as a camouflage and as a means to keep their bodies warm in a forest environment.

As an addendum, fake tans that come out of a bottle play little or no photo-protective role; they can be viewed as either paints that colour skin or chemicals that alter the colour of endogenous substances such as keratins.

8.2.3 Action spectrum for delayed tanning

The action spectrum for delayed tanning corresponds broadly with the erythema action spectrum. The threshold doses at all wavelengths for erythema and pigmentation are similar for poor tanners (skin types I and II), whereas in those subjects who are genetically capable of tanning easily (skin types III and IV), the melanogenic doses in the UVA region are approximately one quarter of the doses required to produce a minimal erythema.

8.2.4 Sun sensitivity and melanogenesis

We saw from figure 4.8 that the Fitzpatrick phototypes correlate poorly with measured minimal erythema dose (MED). In an attempt to simplify classification of UV sensitivity of human skin, a concept based on three melano-categories has been suggested [1]. These categories are:

- *Melano-compromised*—This group takes care not to expose their skin too long without any protection as they have the sort of skin that goes red easily. This redness can last several days and the skin sometimes peels. They experience difficulty getting a tan.
- *Melano-competent*—People in this group do not worry too much about protecting their skin unless they intend being outdoors for a long time. Their skin tolerates the Sun well and they usually tan, especially after a few days of sun exposure.
- *Melano-protected*—Sun exposure is not a problem for this group as they were born with brown/black skin and rarely experience redness from staying in the Sun too long.

8.3 Epidermal hyperplasia

In addition to tanning, the skin is capable of another, perhaps even more important adaptive response, which limits damage from further ultraviolet exposure—epidermal thickening or hyperplasia. This begins to occur around 72 h after exposure, is a result of an increased rate of division of basal epidermal cells, and results eventually in thickening of both epidermis and stratum corneum that persist for several weeks.

This adaptive process, unlike tanning, occurs with all skin types, and is the major factor that protects those who tan poorly in sunlight, i.e. skin types I and II. Even people who are unable to tan, such as those with the skin disease vitiligo, will adapt to repeated sun exposure.

That epidermal hyperplasia occurs mainly following UVB exposure, rather than UVA, is shown by the poor sunburn protection achieved with a UVA-only induced tan, such as that from a sunbed, compared with an equivalent tan achieved from natural sunlight exposure, in which the UVB component of solar UV is the biologically-dominant waveband.

Thickening of the skin, especially of the stratum corneum, after sun exposure can lead to a significant increase in protection against UV radiation by a factor of five or even higher. However, the mechanisms involved in this adaptation appear to be more complex than simple thickening of cell layers, since UV-induced changes in the protein content of the stratum corneum may on their own result in decreased transmission of UV to deeper skin layers.

In white skin, thickening is probably more important than tanning in providing endogenous photoprotection, although in darkly pigmented races it is likely that skin pigmentation is the most important means of protection against solar UV radiation.

8.4 Damage repair

Despite the fact that people are exposed to potentially harmful levels of solar UV radiation, mechanisms have evolved to protect cells and to repair damaged molecules. The cell component most vulnerable to injury is nuclear DNA.

8.4.1 DNA repair

DNA damage occurs spontaneously following UV exposure and cells have developed DNA repair mechanisms to remove damage, although these may not always be fully effective. At its most extreme, the burden of DNA damage may be too great for the cell, either because of its magnitude or because it blocks a vital process and the cell will die in a well-controlled way—a process known as apoptosis. Alternatively, the cell may be able to recognise and eliminate the sites of damage, thus restoring the original DNA sequence. Such repair mechanisms are sometimes said to be error-free, in order to distinguish them from those that allow the cell to eliminate the lesion, but that do not necessarily restore the original DNA sequence and so introduce possible changes, known as mutations, into its genetic code.

Cells are considerably more sensitive to DNA damage at specific stages of the growth cycle but in order to avoid these consequences, cells possess mechanisms whereby the presence of DNA damage activates a signalling pathway involving the tumour suppressor gene p53 (see section 8.4.2) that induces cell cycle arrest prior to the onset of DNA replication.

The main pathways of DNA repair in human skin are:

- *Excision repair*, which involves a variety of proteins that recognise damaged nucleotides, incise surrounding DNA and remove the DNA fragment containing the photodamage;
- *Post-replication*, or recombinational, repair in which UV-damaged DNA can replicate in such a way that gaps are left in the daughter strand opposite the damaged sites, with subsequent filling of the gaps by DNA synthesis.

There is a third pathway of DNA repair referred to as *photoreactivation*, or photoenzymatic, repair in which an enzyme (photolyase) binds to specific photo-lesions such as cyclobutane pyrimidine dimers, and after absorption of visible light reverses the damage *in situ*. However, it is generally accepted that this process does not occur in human skin, only in non-placental mammals.

8.4.2 The p53 tumour suppression gene

The tumour suppression gene p53 is very important for cells in multi-cellular organisms to suppress cancer, and mutation of p53 is a frequent event in skin cancer. P53 has been described as 'the guardian of the genome' for its role in preventing the accumulation of genetic alterations that are seen in cancer cells. The gene plays an important role in controlling cell cycle and apoptosis (cell death). Defective p53 can allow abnormal cells to proliferate, resulting in cancer, and as many as 50% of all human tumours contain p53 mutants.

In normal cells, the p53 protein level is low. Following UV exposure, DNA damage may trigger the increase of p53 proteins, which can lead to growth arrest that stops the progression of the cell cycle and prevents replication of damaged DNA. Furthermore, during growth arrest, p53 may activate the transcription of proteins involved in DNA repair. When DNA cannot be repaired, cells go into apoptosis, which can be regarded as a last resort to avoid proliferation of cells containing abnormal DNA.

8.5 Chemoprevention

Conventional photoprotection by sunscreens is entirely prophylactic, i.e. 'passive' photoprotection, and of no value once DNA damage has occurred.

Dietary agents have been investigated for their potential chemopreventive effects and are claimed to provide low to modest 'active' photoprotection at a topical and systemic level. The benefits of using dietary agents for chemoprevention are that they tend to be non-toxic and easily available and may target multiple pathways involved in tumour development and progression. These agents have been found to work by modulating cell signalling pathways, inducing cell cycle arrest, stimulating DNA repair, acting as antioxidants and inhibiting inflammation, and modulation of the immune system.

Examples of chemopreventive compounds include: natural anti-oxidants, e.g., vitamin A; nicotinamide, a form of vitamin B3; resveratrate, found in red wine, grapes, plums and peanuts; flavonoids; lycopene, found in tomato paste; green tea polyphenols; and liposomes containing natural endonucleases or photolyases.

Dietary intervention may have a role at a population level and a reduction in low-fat diet has shown some promise in both animal and human studies in reducing the appearance of solar keratoses and non-melanoma skin cancer. Similarly, supple-mentation of diet with omega-3 fatty acids has been found to lead to a reduction in markers of UV-induced tissue damage.

It would be fair to say, however, that none of these agents provides anything more than modest photoprotection and are certainly no substitute for established methods such as shade, clothing and sunscreens.

Reference

[1] Fitzpatrick T B and Bolognia J L 1995 Human melanin pigmentation: role in pathogenesis of cutaneous melanoma *Melanin: its Role in Human Photoprotection* ed L Zeise, M R Chedekel and T B Fitzpatrick (Overland Park: Valdenmar Publishing Company) 177–82

Chapter 9

Treating the damage caused by solar UV radiation exposure

In the final chapter, we discuss approaches to treating damage to the skin resulting from excessive solar UV radiation, when attempts to limit exposure have been inadequate.

9.1 Treating sunburn

General advice for treating sunburn is to cool the skin by sponging it with lukewarm water or by having a cool shower or bath. For mild sunburn, a cooling lotion, such as calamine lotion, or after-sun cream can be applied. This can help to cool the skin as well as moisturising and relieving the feeling of tightness.

Potent topical corticosteroids, such as Betnovate®, may reduce inflammation if used within a few hours of exposure. For adults, painkillers such as oral aspirin may help relieve pain. Mild to moderate sunburn will resolve after a few days even if not treated, although in severe cases of sunburn accompanied by pain and blistering, healing will take longer.

9.2 Treatment of non-melanoma skin cancer

Whilst non-melanoma skin cancers (NMSC) are the least serious types of human cancer, they should be treated promptly to prevent localised complications such as ulceration, bleeding and infection.

9.2.1 Surgical excision

The treatment of choice for most basal cell carcinomas (BCC) and squamous cell carcinomas (SCC) is surgical excision. A margin of a few millimetres is allowed when removing the tumour to ensure that the malignant cells are completely excised (figure 9.1).

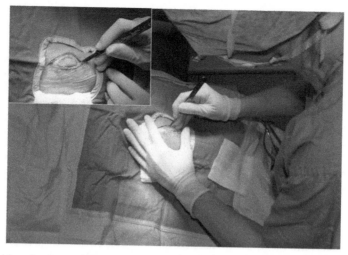

Figure 9.1. Excision of squamous cell carcinoma from the right arm. The inset in the top left corner shows the excision in closer detail (courtesy of Dr James Langtry).

9.2.2 Curettage and diathermy

This technique involves scraping away the tumour with an instrument called a curette (figure 9.2), after which the wound is allowed to heal from the base up in much the same way a graze on the knee may heal following mild to moderate trauma.

The wound is cauterised using diathermy post-curettage to stop bleeding and destroy any tumour that has not been scraped off.

9.2.3 Cryotherapy

Cryotherapy destroys tissue due to the formation of ice crystals in cells that have been exposed to liquid nitrogen at a temperature of −180 °C and is best suited to superficial BCCs and some types of actinic keratoses.

9.2.4 Topical chemotherapy

There are now a range of topical chemotherapeutic agents that are useful in the management of actinic keratoses with generally only a limited role in the treatment of NMSCs. These include 5% 5-fluorouracil cream (marketed as Efudix® in the UK), Diclofenac sodium gel (marketed as Solaraze® in the UK), and Imiquimod cream (marketed as Aldara® or Zyclara® in the UK).

9.2.5 Photodynamic therapy

Topical photodynamic therapy (PDT) involves photoactivation of a tissue-localised photosensitiser, mainly porphyrin agents. PDT is effective for superficial NMSCs and for actinic keratoses, and is especially beneficial for patients with large and multiple lesions and difficult treatment sites such as the lower leg. Topical PDT is generally well tolerated, although pain during irradiation can occur.

Figure 9.2. Surgical instruments used in excision of skin tumours: scalpel (left), 8 mm biopsy punch (middle), ring curette (right) (courtesy of Professor Jonathan Rees, University of Edinburgh).

9.2.6 Mohs micrographic surgery

Mohs micrographic surgery is a specialist type of skin surgery used when a skin cancer has spread beyond its visible margins. It can be useful on sites such as the nose or close to the eye where the margin of the tumour is difficult, or impossible, to define clinically.

The technique is demanding and time-consuming as after excision, the removed tissue is examined histologically and if any residual cancer remains, another excision is carried out. This process is repeated until no malignant cells remain in any of the removed skin tissue.

9.2.7 Radiotherapy

Radiotherapy used to be widely employed in the treatment of NMSCs but its use today is limited to the treatment of aggressive SCCs.

9.3 Treatment of malignant melanoma

The treatment of malignant melanoma is excision under local anaesthetic. Where possible, lesions should be completely excised together with a small amount of normal looking skin (~2 mm) from around the margin.

The sample removed is then examined by a histopathologist, and a further excision carried out based on the Breslow thickness (section 4.4.1), which might mean the removal of more tissue (5–20 mm) from around the first excision.

9.3.1 Prognosis and treatment of advanced disease

The Breslow thickness provides the single best prognostic information for patients who do not have metastases when they first present. If the Breslow thickness indicates that the melanoma may have metastasised, diagnostic tests, that can include sentinel node biopsy, a CT and/or PET scan, and molecular testing, may be performed to confirm or otherwise the presence of metastatic disease.

If metastases are confirmed, the prognosis is poor as there is currently no curative treatment for metastatic malignant melanoma. Treatment of advanced melanoma may involve surgery, radiotherapy, chemotherapy, immunotherapy and targeted therapy aimed at patients who have a mutation in the BRAF gene.

9.3.2 On-going surveillance

The goal of any cancer follow-up regimen is to identify recurrence or metastasis early and to initiate treatment in the hope of having a positive impact on the long-term outcome. Melanoma is no exception and patients who have had the disease have a 4%–8% lifetime risk of developing a secondary primary melanoma.

General guidelines for on-going care in patients who have been treated for melanoma include clinical evaluation two to four times a year, at least initially. The frequency of evaluation will depend upon several factors that include tumour stage, history of multiple melanomas, presence of atypical nevi, family history of melanoma, patient anxiety, and the patient's ability to recognise signs and symptoms of disease. Patients should also be educated to perform monthly examination of their skin and lymph nodes. Routine on-going laboratory tests and imaging studies are normally not indicated in asymptomatic patients.

9.4 Treatment of photoaged skin

In today's world there is more demand than ever to maintain a youthful look. This, coupled with an ageing, but often physically fit and economically robust, population has led to a plethora of techniques available to mitigate the effects of photoageing on the skin.

Photoageing occurs as the result of chronic sun exposure but the extent and consequences of these changes vary greatly among individuals. The decision about whether to seek treatment for photoaged skin depends on the nature of the changes, their severity, the degree to which they impact on a subject's self-image, and his/her willingness to accept the risks and costs of treatment. Treatment can be topical preparations, minimally-invasive procedures, or invasive surgery.

9.4.1 Topical preparations

Alpha-hydroxy acids
Alpha-hydroxy acids (AHA) are compounds derived from dairy products, fruit or sugar cane. Topical treatment of photodamaged skin with AHA has been reported to improve wrinkling, roughness and mottled pigmentation within months of daily application. AHA in low concentrations (typically 4%–12%) are common components of cosmetic creams and lotions that are promoted as contributing towards the amelioration of ageing of skin.

Topical retinoids
Topical retinoids, which are vitamin A derivatives, can result in modest improvement in fine wrinkling, mottled pigmentation, sallowness, and skin roughness.

9.4.2 Minimally-invasive procedures

There are many procedures, ranging from relatively non-invasive approaches to those that are quite invasive, that are used for rejuvenation of (mostly facial) photoaged skin. For most procedures, the durability of beneficial effects is usually measured in months or a few years, and the risks can include bruising, infection, scarring, pain, pigment change, and prolonged healing. Some of the technologies currently in use include the following.

Ablative non-fractionated lasers
The word *ablation* stems from the Latin word *ablatus* meaning 'carried away' and ablative skin resurfacing removes the epidermis and superficial dermis and produces the most dramatic laser-treated results for skin resurfacing.

In ablative laser surgery, the infrared radiation super-heats cellular water leading to vaporisation of skin cells in a skin-peeling effect. This effect promotes collagen formation and retraction of the dermis and epidermis to tighten the skin resulting in an improvement in photodamaged skin. The greater the depth of injury of the dermis, the greater the degree of skin resurfacing and the greater the degree of adverse effects and healing time.

The two lasers commonly used both emit infrared radiation and are the carbon dioxide laser (10.6 μm) and the erbium-doped yttrium aluminium garnet (Er:YAG) laser (2.94 μm).

Non-ablative non-fractionated lasers
Whereas ablative laser surgery vaporises the tissue, non-ablative laser surgery simply coagulates the affected tissue. Non-ablative lasers produce a gentler effect on the skin, inducing controlled tissue injury in the dermis, which stimulates dermal remodelling and collagen production. The potential damaging risks associated with non-ablative lasers are significantly lower compared with ablative lasers. Again, these lasers emit infrared radiation and examples include the Nd:YAG laser (1.32 μm) and the diode laser (1.45 μm).

Ablative fractionated lasers
The most recent generation of ablative lasers are the fractionated ablative lasers, either CO_2 or Er:YAG, that have been around since 2007. They are safer than their non-fractionated counterparts but still retain a high risk of potential damage in the form of scarring, discoloration, and skin infection. Their main use is for mild skin tightening to combat laxity as well as the treatment of photodamage.

Non-ablative fractionated lasers
Non-ablative fractionated lasers combine the best of the gentle and safe aspects of both fractionated and non-ablative technologies, and entered the market in the mid-2000s. This class of laser, which includes the Nd:YAG laser (1.44 μm) and erbium glass lasers (1.55 μm), is aimed towards improving texture, mild to moderate wrinkles, and pigmentary changes due to sun damage.

In general, ablative technologies offer improved cosmetic results but at the cost of longer recovery times and potentially more severe side effects. Non-ablative technologies, on the other hand, usually offer more moderate results with fewer side effects and an easier recovery. Fractionated technologies appear to combine the best aspects of each category with shorter recovery times and results approaching those of fully ablative technologies. An example of fractionated laser resurfacing is shown in figure 9.3

Radiofrequency
Radiofrequency (RF) systems thermally heat tissue with the RF radiation penetrating more deeply than laser radiation. Much like the laser systems, however, RF systems achieve results by denaturing existing collagen and stimulating production of newer and shorter collagen, leading to lasting tissue tightening.

Intense pulsed light
Intense pulsed light (IPL) devices emit incoherent, high intensity light usually in the spectral interval 500 nm to 1200 nm.

The principle of the treatment is selective photo-thermolysis, in which thermally mediated radiation damage is confined to selected epidermal and/or dermal chromophores, either haemoglobin, water, or melanin, at the cellular or tissue levels. By optically filtering the light output, the device can be targeted to specific applications. For example, green light is optimal for vascular and pigmented lesions such as sunspots, whereas orange/red light is selected for epilation and treating scars.

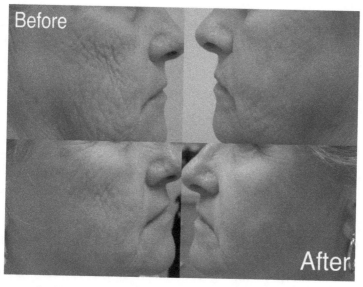

Figure 9.3. An example of laser resurfacing on the face (courtesy of Dr Pillay, The Wembley MediSpa, Cape Town).

Intense pulsed light technology is claimed to be a safe and effective modality not only for the treatment of vascular and pigmented lesions, but also epidermal and dermal atrophy associated with photoageing.

Dermabrasion
Dermabrasion is a procedure that uses a rapidly rotating device to resurface the epidermis and upper dermis by means of mechanical abrasion. Dermabrasion can decrease the appearance of fine facial lines and improve the look of sun-damaged skin. After dermabrasion, the skin that grows back is usually smoother and younger looking.

Botulinum toxin type A (Botox)
Botox injections are the best known of a group of medications that use various forms of *Botulinum* toxin to temporarily paralyze muscle activity. By means of neurotoxic effects, Botox reduces the tone of the muscles. The most common use of these injections is to temporarily relax the facial muscles that underlie and cause wrinkles, such as frown lines between the eyebrows, lines that fan out from the corners of the eyes (crow's-feet) and forehead furrows. The procedure is not without risk and possible, but rare, side effects include pain, swelling or bruising at the injection site, headache or flu-like symptoms, droopy eyelid, crooked smile or drooling, and eye dryness or excessive tearing.

In addition to their cosmetic applications, Botox injections are also used to treat such problems as cervical dystonia (repetitive neck spasms) and hyperhidrosis (excessive sweating). Botox injections may also help prevent chronic migraines in some people.

Dermal fillers
These are injections used to fill out wrinkles and creases in the skin. They can also be used to increase the volume and definition of the lips and cheeks. The fillers are made from a variety of materials and the effects can be either temporary or permanent, depending on the type of filler, which include collagen (bovine, human or autologous), hyaluronic acid, and synthetic fillers such as methylmethacrylate and polyacrylamide gel.

Chemical peels
Chemical peels are among the oldest technologies of skin rejuvenation and are classified as superficial, medium or deep based on the depth of the procedure. Superficial peels produce exfoliation of the epidermis and may be achieved by a wide variety of agents that include AHA in concentrations generally between 20%–70%, and trichloroacetic acid (TCA; 10%–30%). Deeper peels use more caustic agents such as phenol that penetrate the reticular dermis with the aim of correcting the most severe photodamage.

Peels can be an effective treatment for facial blemishes, wrinkles, and uneven skin pigmentation.

9.4.3 Invasive surgery

Invasive surgery for facial photodamaged skin, better known as a face-lift, is a procedure that is typically used to give a more youthful appearance to the face. This type of cosmetic surgery works best for the lower half of the face, and particularly the jaw line and neck. Its aim is to remove excess facial skin but may also include tightening of underlying tissues. Features of a photodamaged skin such as laxity with fine wrinkles, freckles and rough areas will benefit more by chemical peel or laser resurfacing.

The best candidate for a face-lift is someone whose face and neck has begun to sag, but whose skin still has some elasticity and whose bone structure is strong and well defined. Most patients are in their 40s to 60s, but facelifts can be done successfully on people in their 70s or 80s.

CPSIA information can be obtained
at www.ICGtesting.com
Printed in the USA
BVHW02*1735120118
504764BV00002B/2/P